Green Energy and Technology

More information about this series at http://www.springer.com/series/8059

Harry Apostoleris · Marco Stefancich
Matteo Chiesa

Concentrating Photovoltaics (CPV): The Path Ahead

Springer

Harry Apostoleris
Laboratory for Energy and Nano Science
Khalifa University of Science and Technology, Masdar Campus
Abu Dhabi
United Arab Emirates

Matteo Chiesa
Laboratory for Energy and Nano Science
Khalifa University of Science and Technology, Masdar Campus
Abu Dhabi
United Arab Emirates

Marco Stefancich
Dubai Electricty and Water Authority
Dubai
United Arab Emirates

ISSN 1865-3529 ISSN 1865-3537 (electronic)
Green Energy and Technology
ISBN 978-3-319-62979-7 ISBN 978-3-319-62980-3 (eBook)
DOI 10.1007/978-3-319-62980-3

Library of Congress Control Number: 2017951179

Printed on acid-free paper

This Springer imprint is published by Springer Nature
The registered company is Springer International Publishing AG
The registered company address is: Gewerbestrasse 11, 6330 Cham, Switzerland

Acknowledgements

We acknowledge the contribution of Maritsa Kissamitaki for production of the figures; Ibraheem Almansouri for helpful discussion; and the Abu Dhabi Education Council for financial support of this work.

Contents

Chapter 1
What Went Wrong with CPV?

For decades, green-energy advocates and solar power researchers struggled with a simple problem: nearly endless energy, freely available from the sun, and no economical way to collect it. With solar cell costs higher by an order of magnitude than conventional energy generation at the turn of this century, a straightforward solution presented itself: get a single solar cell to generate a large amount of energy by using cheap optical devices to funnel light from a large area into a small cell. By circumventing the cost issue in this way, these "concentrator photovoltaics," or CPV, were hailed as the most promising pathway to low-cost solar electricity.

Since then, CPV has fallen on hard times. The photovoltaic sector overall has boomed in this decade, with tens of gigawatts of new capacity installed each year [1], but CPV has almost disappeared as an industry player [2]. As recently as a few years ago, industry leaders saw CPV as a potentially sector-leading technology, offering high power densities and low costs. Now the industry has fled and most CPV companies have shut down or moved on to other technologies. What happened to CPV, and does it offer any lessons or experience relevant to the now-booming PV sector?

The numbers are difficult to argue with. Originally conceived as a cost-saving technology, it is increasingly clear that CPV has been beaten at its own game, undercut by plummeting silicon prices and declining PV cell manufacturing costs. It certainly appears that CPV has no real place in the market, while conventional Silicon-based PV has become one of the most affordable energy-generation technologies across the world. CPV has lost, or at least is badly losing, the "race" to take solar energy mainstream. Is this the end of the line, or can it reinvent itself to become relevant and useful again in this new solar-powered age?

In this book we aim to provide a context and a perspective on the history and potential future of CPV a bit different from that of other authors on the subject. Rather that focus on its current technological state, which has been well-covered [3–7], we will seek to provide a narrative of the development of CPV and its constituent components, telling the story of *how* the technology reached the point where it stands today, and what pathways forward may exist based on new

H. Apostoleris et al., *Concentrating Photovoltaics (CPV): The Path Ahead*, Green Energy and Technology, DOI 10.1007/978-3-319-62980-3_1

innovations. We will consider emerging developments in the main technological areas essential to CPV, to see what elements can be put together in new ways to create a viable future.

We will try to describe the ways in which CPV, or technologies associated with it, can continue to be useful in this time when directly undercutting the cost of conventional PV seems out of the question.

But before doing that we will begin to address a question that will guide much of this work: what went so wrong with CPV?

1.1 Why Is CPV Losing the Race?

Until a few years ago, CPV was a talked-about technology seen as a strong competitor to standard, silicon-based solar panels [2, 8]. Why did interest in CPV dry up so suddenly? It's all about the money.

A decade ago, when interest in CPV was at its high point, the economics of PV were much different than they are today. The price of silicon-based PV panels was several times higher [9], trapping solar in the category of "alternative energy," favored by environmentalists but unable to compete on cost with coal or gas and heavily dependent on government subsidies. Most of the manufacturing cost of the module came from the cost of PV cells themselves; high cell costs resulted both from the smaller scale and lack of manufacturing experience, and from a bottleneck in the production of PV-quality silicon which at its peak drove raw material prices to 30 times their current levels. The clearest path to closing the cost gap with fossil fuels seemed to be reducing the amount of PV cell used. Simple schemes were developed to enhance PV panel with mirrors, and concentrator modules were developed using nearly every optic imaginable combined with high quality Silicon cells [3]. One of these many companies was in fact founded by one of the authors of this book [10]. As a result we have had a front-row seat to what happened over the next several years.

In 2008 the silicon shortage ended as new production capacity came on line, and the combination of plummeting material costs and continued scaling of manufacturing began to transform PV economics [11]. Over the next few years, the costs of Si cells became low enough that concentration became less and less attractive as a cost-cutting approach, and nearly all of the silicon-based CPV manufacturers (the author's company sadly included!) failed. There was, however, another application of the technology which persisted a bit longer. At the same time as Silicon concentrators were being developed, researchers saw that concentration offered a work-around to the very low efficiency of commercial silicon PV cells of the time. High-efficiency multijunction solar cells originally developed for spacecraft offered a solution to the efficiency problem [12]. These cells were and remain much more expensive than Si cells, but could more than double Si cell efficiencies. The approach of cutting costs by reducing cell area via concentration gave these high-efficiency, multi-junction space solar cells a new life on earth [13].

For some years—in fact until quite recently—it seemed that the advantage of high conversion efficiency might give this conception of CPV and edge over standard "flat-plate" PV [14]. But the last few years have shown fairly definitively that this, too, is not to be. Installations of high-efficiency CPV peaked in 2012 with over 120 MW installed [15]. But in the years after, demand plummeted and most CPV companies failed.

Why did this happen? Answering this question is the purpose of the first part of this book. We will tackle this question in two parts. First we will survey the PV industry at large to get as better picture of what any competing technology has been up against over the last ten years. Then we will dig into the technological paradigm that dominates today's CPV industry, to see why it was not up to the task.

1.2 Status Update on PV

Based on a quick survey of the PV industry's development in this decade, it seems likely that future generations will look back on this decade as a major turning point in the energy sector, driven by the widespread adoption of low-cost PV. A few key figures tell the story:

1.2.1 PV Is Growing—Fast

At first glance solar energy may not seem like a major energy player—it still accounts for less than 0.5% of total global energy consumption, and just over 1% of electricity consumption [1]. But a closer look reveals is strength. Based on figures from the International Energy Agency, the total installed PV capacity in the world has grown from 40 GW in 2010 to nearly 240 GW in 2015 [16]. The transition has been highly visible on rooftops, as homeowners install small residential systems, but just as transformative have been the rise of large 'utility-scale' solar parks that sell their power through the electric grid. The cause of this growth? A precipitous drop in the cost of PV which has accelerated during this decade, causing more and more consumers and producers to see PV as a money-saving choice.

1.2.2 PV Is Cheap

From 2010 to the beginning of 2016, the cost of PV panels dropped from \$1.9/W to \$0.57/W in January 2016 [9]. We can put this into a more familiar context by looking at Power Purchase Agreements (PPAs), the contracts that utility companies sign with providers of electricity. In the US, the range of contracted prices for solar energy has gone from 10–15 c/kWh in 2010 down to the 3–6 c/kWh range at the beginning of 2016 [17]. In 2015, responding to a deal by Nevada's public utility to buy solar electricity for 3.87 c/kWh, an observer for Bloomberg called it "Probably the cheapest PPA I've ever seen in the US." [18] And these were not even the world's lowest prices for solar electricity—Dubai's Al Maktoum solar park will sell energy to the grid at 2.99 c/kWh, which beats the cost of coal and gas by a substantial margin. For comparison, The price paid for electricity by a residential consumer in the US was about 12 c/kWh in 2016 [19].

1.2.3 The More It Grows, the Cheaper It Gets

What is behind these recent developments? It is possible to point to specific events, such as the end of the silicon shortage in 2008, recent oversupply of PV modules, and low interest rates that ease capital investment, that contribute to low prices and increased consumer interest. But if we again take the long view, we can put recent events into a broader, and very encouraging, context. It was first noted by Richard Swanson, the founder of PV manufacturer SunPower, in 2006, that the prices of photovoltaic modules follow a learning curve typical of mass produced goods: prices decrease as more pieces are produced [20]. It is this, more than any short-term trend, that indicates the likelihood of not only a bright, but a nearly inevitably dominant future for photovoltaics, as the economies of scale drive prices ever lower and feeds a cycle of cost reductions and increasing demand. There is only one main conclusion that must be updated from Swanson's analysis: in his paper he predicted that module prices would reach $0.65 by 2023, a prediction which came true—in 2013! [9].

A caveat is that the incredible recent decline in prices may be the result of intentional oversupply by certain market players, whose goal is to eliminate their competition and dominate the industry. To the extent that this is true, it seems likely that prices will stabilize as market consolidation sets in. But in the long term the best bet is that costs will continue their gradual decline and PV will become cheaper and cheaper as time goes on and the industry gets bigger. And the biggest takeaway is that, while many observers remain mired in the past, believing PV is a promising technology that requires more work to become affordable, exactly the opposite is true. PV is already, today, so cheap that it is better to be asking whether conventional energy technologies have what it takes to compete in a free market with solar! (Fig. 1.1).

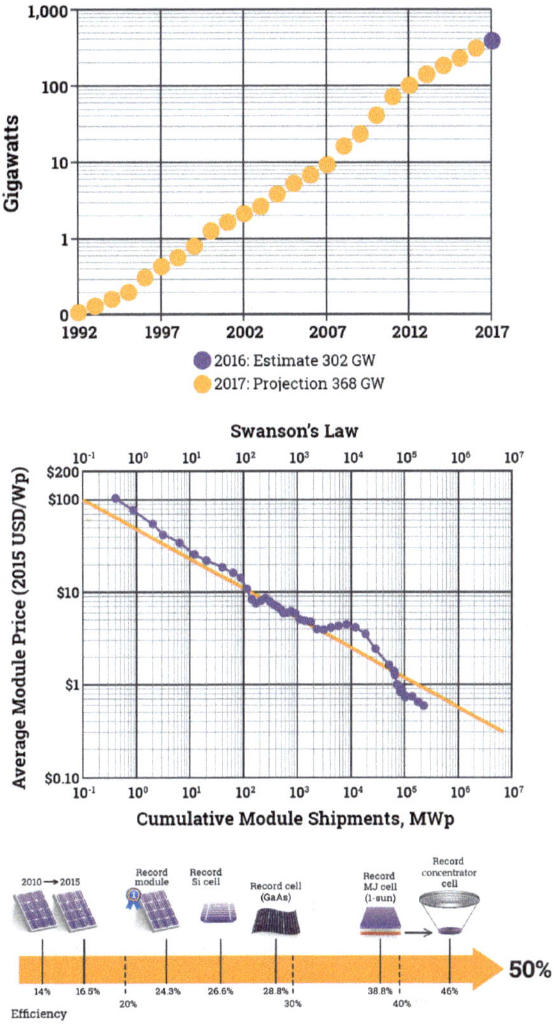

Fig. 1.1 Key figures on the growth of photovoltaics. *From top* total installed PV capacity [21]; learning curve for photovoltaics manufacturing, relating price decline to production volume [23]; Efficiency landmarks for PV technologies

1.3 Improving PV System Performance

The three figures above paint the general picture of the market status of PV: the growing size of the market, declining costs, and the "virtuous cycle" (the learning curve) by which these two developments compound each other. The big takeaway is that solar power may already be, and certainly will be in the near future, the cheapest way to produce electricity: it is not an 'alternative energy' but a fully mainstream technology. As with any commercial technology, the money will be in making the best-performing and most consumer-friendly version possible. To assess the performance trends of PV, we can track the improvements in module efficiency—how effectively the panel converts light into electrical power.

According to a study at Lawrence Berkeley National Labs, commercial PV module efficiencies average about 17% [17]. This is where the cost-per-watt is optimized (for now), but it is far from the best that a solar panel can do. In 2016 the record for a panel efficiency was 24.4% (after being broken four times in the preceding two years); the cell efficiency record climbed from 25.6 to 26.6% [24]. While panel efficiencies are improving quickly (2016 saw the previous record of 22.5% broken three times), cell efficiencies are climbing much more slowly. The reason for the slow growth is clear: we are approaching the fundamental efficiency limits of silicon. Based simply on the material properties of silicon, the limiting efficiency is calculated to be 29% [25–27]. The realistic limit taking into account inevitable losses due to other material properties was once believed to be 26%, but recent developments have allowed this value to be exceeded [28]. We have more or less reached the end of what can be done, technologically, to improve silicon solar cells, but cost reductions for these high-efficiency products can be expected to continue into the near future which will lead to efficiency improvements in mainstream commercial products of several percent.

This is the starting point from which we can construct the "standard" argument for CPV: that additional benefits can be gained from increasing the efficiency of PV systems, and that therefore we need to look beyond silicon for new cell technologies that will provide higher efficiencies and greater power output. This, it is argued, may in turn lead to further cost reductions as higher output allows the initial cost of the system to be paid back more quickly. The market disagrees with this assessment. To understand why, we turn to the details of CPV technology, and show why it comes up short of meeting its lofty expectations.

References

1. BP, Statistical Review of World Energy. (2016) London: British Petroleum.
2. Kurtz, S. (2009). *Opportunities and challenges for development of a mature concentrating photovoltaic power industry*. National Renewable Energy Laboratory.
3. McConnell, R. (2005). Concentrator photovoltaic technologies: Review and market prospects. *Refocus, 6*, 35–39. doi:10.1016/S1471-0846(05)70429-7.

4. Philipps, S., Bett, A., Horowitz, K., & Kurtz, S. (2015). *Current status of concentrator photovoltaic (CPV) technology*. Golden, CO: National Renewable Energy Laboratory (NREL).
5. Shanks, K., Senthilarasu, S., & Mallick, T. K. (2016). Optics for concentrating photovoltaics: Trends, limits and opportunities for materials and design. *Renewable and Sustainable Energy Reviews, 60,* 394–407.
6. Algora, C., & Rey-Stolle, I. (2016). *Handbook on concentrator photovoltaic technology*. New York: Wiley.
7. Khamooshi, M. et al. (2014). A review of solar photovoltaic concentrators. *International Journal of Photoenergy,* 2014.
8. Swanson, R. M. (2000). The promise of concentrators. *Progress in Photovoltaics: Research and Applications, 8,* 93–111.
9. Barbose, G. L., & Darghouth, N. R. (2016). Tracking the Sun IX: The Installed Price of Residential and Non-Residential Photovalic Systems in the United States. Washington: US Department of Energy.
10. Antonini, A., et al. (2009). Rondine® PV concentrators: Field results and developments. *Progress in Photovoltaics: Research and Applications, 17,* 451–459. doi:10.1002/pip.907.
11. Shah, V., & Booream-Phelps, J. (2015). *Crossing the chasm solar grid parity in a low oil price era*. Deutsche Bank, February.
12. Yamaguchi, M. (2003). III–V compound multi-junction solar cells: Present and future. *Solar Energy Materials and Solar Cells, 75,* 261–269.
13. Cotal, H., et al. (2009). III–V multijunction solar cells for concentrating photovoltaics. *Energy & Environmental Science, 2,* 174–192.
14. Haysom, J. E., Jafarieh, O., Anis, H., Hinzer, K., & Wright, D. (2014). Learning curve analysis of concentrated photovoltaic systems. *Progress in photovoltaics: Research and applications*.
15. Photovoltaics Report. (2015). Fraunhofer-Institut für Solare Energiesysteme ISE.
16. Snapshot of Global Photovoltaic Markets 1992–2015. (2016). *International Energy Agency IEA*.
17. Barbose, G. L., & Darghouth, N. R. (2016). Tracking the Sun IX: The installed price of residential and non-residential photovalic systems in the United States. Washington: US Department of Energy. Available at emp. lbl. gov/publications/trackin.
18. Clover, I. (2015). PV Magazine.
19. Administration, U. S. E. I.
20. Swanson, R. M. (2006). A vision for crystalline silicon photovoltaics. *Progress in Photovoltaics: Research and Applications, 14,* 443–453. doi:10.1002/pip.709.
21. Growth of Photovoltaics. https://commons.wikimedia.org/wiki/File:PV_cume_semi_log_chart_2014_estimate.svg.
22. Trube, J., Fischer, M., & Metz, A. (2016). Wiley-VCH Verlag GmbH Postfach 101161, 69451 Weinheim, Germany.
23. Green, M. A., et al. (2017). Solar cell efficiency tables (version 49). *Progress in Photovoltaics: Research and Applications, 25,* 3–13. doi:10.1002/pip.2855.
24. Kerr, M. J., Cuevas, A., & Campbell, P. (2003). Limiting efficiency of crystalline silicon solar cells due to Coulomb-enhanced Auger recombination. *Progress in Photovoltaics: Research and Applications, 11,* 97–104.
25. Tiedje, T., Yablonovitch, E., Cody, G. D., & Brooks, B. G. (1984). Limiting efficiency of silicon solar cells. *IEEE Transactions on Electron Devices, 31,* 711–716.
26. Richter, A., Hermle, M., & Glunz, S. W. (2013). Reassessment of the limiting efficiency for crystalline silicon solar cells. *IEEE Journal of Photovoltaics, 3,* 1184–1191.
27. Swanson, R. M. (2005). *Photovoltaic Specialists Conference, 2005. Conference Record of the Thirty-first IEEE.* IEEE. pp. 889–894.

Chapter 2
The Case for CPV

The energy-generating potential of photovoltaics is huge—but it does suffer from some practical challenges, particularly related to the required size of generating installations. The solar resource is quite dilute, which means that a photovoltaic power plant needs to occupy a very large area compared to conventional power plants to generate a given amount of output. The amount of space required depends inversely on the efficiency of the solar panels—so a power plant of say, 100 MW that uses panels with 20% efficiency will occupy 25% less space than a plant of the same capacity using panels of 15% efficiency.

As we saw in the last chapter, Si technology, while extremely low in price, is nearing the physical limits of its efficiency potential. CPV has tried to circumvent that by enabling the use of much higher efficiency solar cells which cost a hundred or more times as much as Si cells. Any thorough discussion of CPV needs to start with an understanding of these cells, their operating mechanism and limits, and the forces that led to their development.

2.1 Operating Principles and Limits of Solar Cells

Photovoltaic cells work by absorbing the energy of solar photons to excite electrons in a semiconductor (for in-depth discussion, see, for example, the texts by Nelson or Green on solar cell physics [1, 2]). The basic mechanism is illustrated in Fig. 2.1. A basic solar cell is an extremely simple device—a p-n junction. One of the two main layers, usually the p-type, is much thicker than the other and acts as the "absorber." Incident photons travel through the absorber until they are absorbed by an interaction with an electron. When an electron absorbs a photon, it is excited from the valence band of the absorber into the conduction band, creating a pair of free charge carriers—the conduction band electron and the valence band hole left in its place. The built-in electric field of the p-n junction then separates the

H. Apostoleris et al., *Concentrating Photovoltaics (CPV): The Path Ahead*,
Green Energy and Technology, DOI 10.1007/978-3-319-62980-3_2

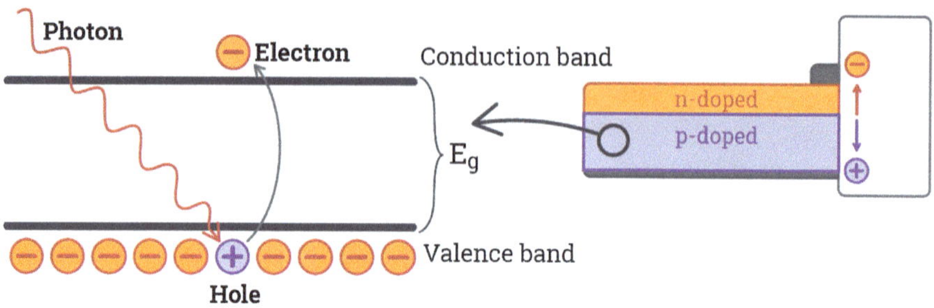

Fig. 2.1 The operating principle of a photovoltaic cell: valence-band electrons in a semiconductor are excited to the conduction band by the energy of incident photons. Conduction band electrons and valence band holes are separated by a PN junction, causing current to flow through the external circuit

Fig. 2.2 Interaction of photons of different energies with a PV cell: photons with energy below the band gap are transmitted; those with energy above the band gap create a carrier pair by promoting an electron, but the difference between the photon energy and the band gap becomes heat

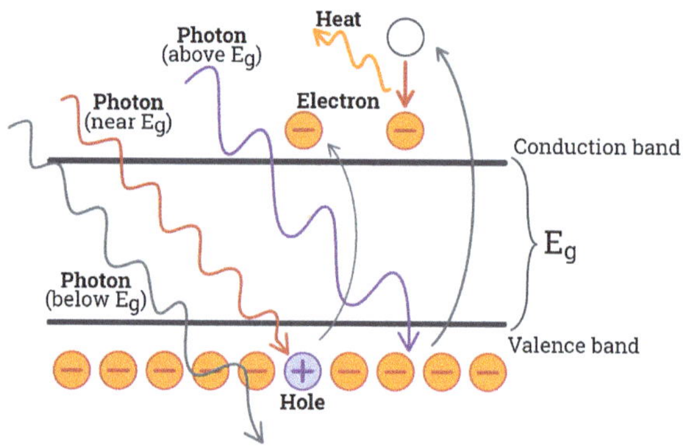

electron-hole pair and drives them into the external circuit, where electrons flow in one direction and holes in the other, producing an electric current.

This extremely basic picture of solar cell operation is sufficient to explain the strongest limits on solar cell efficiency. The critical understanding is how the solar cell interacts with photons of different energies. This is shown in Fig. 2.2. The band gap of the absorber acts as a threshold energy that determines what happens to the energy of an incident photon. First, an electron can only cross the band gap if it absorbs a photon whose energy is greater than the band gap energy E_g. Photons with energy less than E_g do not create carrier pairs, and typically pass through the solar cell entirely to be transmitted out the back side (or absorbed by the back contact). A photon whose energy is higher than E_g can be absorbed, transferring all of its energy to the electron. What happens next also depends on the photon energy. If the photon has energy very close to E_g, the electron is excited to the bottom of the conduction band and essentially all of the photon energy is transferred to the carrier pair. However if the energy is significantly higher than E_g, the electron will be promoted to a higher-energy region of the conduction band. The electron will then seek to minimize its energy by relaxing, very quickly, to a lower-energy state near the conduction band edge. The difference between the photon energy and the band gap energy is dissipated as heat into the solar cell. This energy is not carried by the extracted current and represents a *thermalization loss*.

These two mechanisms—below band gap transmission and above band gap thermalization loss—are the two biggest factors that fundamentally limit the efficiency of an ideal solar cell [3]. There are other fundamental losses, related to recombination of carrier pairs inside the cell, and the mismatch in the entropy carried by light inside and outside of the cell [4]. These will be touched on later, incidentally, but not belabored except where they are directly relevant to the subject under discussion. Because of the combination of these factors, the ultimate limit for a solar cell with a single absorber material—a so-called "single-junction" device—is only 33% at the optimum value of the band gap (about 1.5 eV), and declines further for values of E_g far from this optimum. When accounting for its band gap and additional fundamental loss mechanisms specific to the material, silicon solar cells can be shown, as we noted in the previous chapter, to be limited to 29% efficiency [5]. This value has nearly been reached in the laboratory and the efficiencies of commercial panels continue their climb toward something near these experimental records [6, 7]. In order to go beyond this, different cell technologies are needed.

2.2 Solar Cells to Match the Solar Spectrum

The basic problem of photovoltaic efficiency can be considered in the following way: we are trying to capture a resource with a broad spectrum using a device with a narrow spectral response. Solar photons are distributed over a range of about one order of magnitude in energy. A solar cell, as we have seen, functions best when incident light is as close to its band gap—a single energy—as possible. In order to fully utilize the resource using a photovoltaic converter, we need to broaden its spectral response to match the incoming spectrum of light. This is the concept of so-called "third-generation photovoltaics." [8]

One strategy has come to dominate the push for high-efficiency PV. It grows from the observation from the last section that solar cells are typically transparent to below-E_g radiation. This means that if a second collector—a solar cell, thermal absorber or anything else that might be useful—is added behind the solar cell, the sub-E_g radiation will be collected by this second device rather than the solar cell. This is the key insight that led to the development of the multijunction solar cell. If a number of solar cells are stacked, from top to bottom, in decreasing order of absorber band gap, the incident light will be divided into smaller bands, defined by the band gap energies of the solar cells. Fig. 2.3 illustrates this for a typical combination of band gaps—each cell spontaneously "filters" the above-E_g light for electricity production and transmits the below-E_g to the cell below. This avoids the need for balancing between below-E_g loss and thermalization loss. Multijunction cells offer huge efficiency gains over single-junction: as of this writing the record

Fig. 2.3 Division of the solar spectrum by a multijunction cell

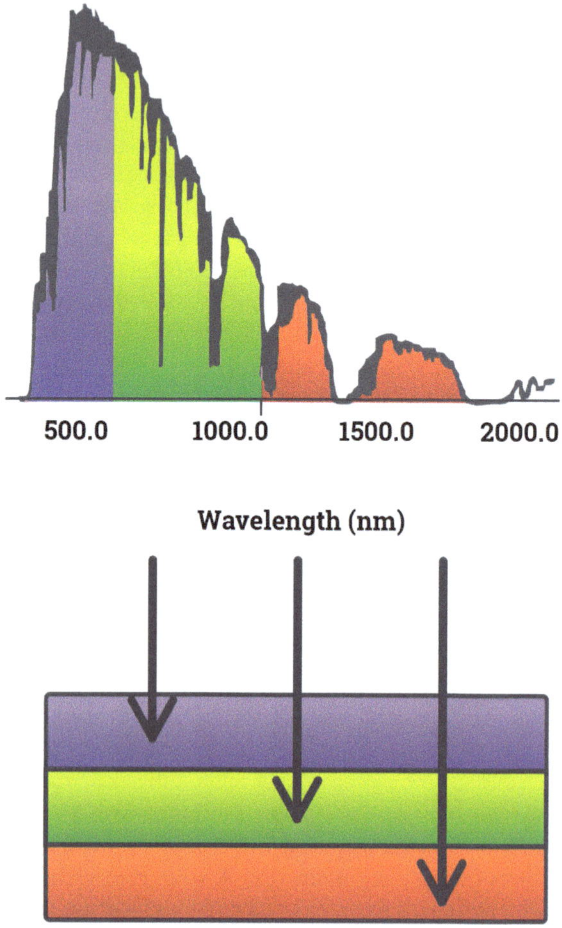

efficiency of a multijunction cell was just below 39% under unconcentrated sunlight [9] and 46% under light concentrated about 300 times [10]. For comparison, the best-ever single-junction cell is "only" 28.8% [11, 12].

The first key to making this kind of cell is to find combinations of photovoltaic materials with the correct band gaps or ideally, a single family of materials whose band gap can be tuned across the desired range by varying their composition. Several of these material families, in fact, exist. The best-performing and most widely used of these is the III-V alloys—combinations of at least one materials from the III column of the periodic table with at least on from the V column—in particular Ga, In, As, and P in various combinations [13, 14]. While these materials have excellent optoelectronic properties and make high-quality solar cells, they are both extremely expensive due to the scarcity of some of these elements, and require advanced processing techniques to make high-quality solar cells [13, 14]. As a result these cells are more expensive than Silicon cells by a factor of a few hundred.

For this reason these cells have only two practical applications—aerospace, where the cost of solar cells is a secondary concern, and CPV, where concentration allows the cost to be offset by reducing the cell area by 500 times or more. But this imposes its own constraints.

At this point it is worth considering in a bit more detail why it is helpful to have high-efficiency PV cells at all.

2.3 Why Do We Care About High Efficiency?

The high efficiency of multijunction cells makes them enticing to both researchers and manufacturers looking to reach the 30–40% efficiency range that is well beyond the reach of silicon. But what, really, is the value of high efficiency? In research, efficiency is the standard by which all cells are measured. In real-world deployment the value is more dubious. Experts rightly warn against being an "efficiency snob" when selecting solar panels, instead putting the emphasis on cost and reliability (as one would with any other consumer product!) [15]. This is an important caveat for researchers as much as consumers, that advises us against such crazy schemes as trying to overturn the whole PV manufacturing sector by, say, switching from Si to some new material, for the sake of gaining a few points of efficiency. So we should be clear, if we do insist on chasing efficiency, regarding what exactly it is good for.

The basic idea of increasing efficiency is "getting more from less." But what are we trying to use less of? There are two main perspectives to take. One is that we want to use less hardware—fewer solar panels, less racking and mounting material, a smaller number of electrical connections, lower expense in terms of time and money to install. Ultimately this is an argument based on **cost**. We try to get the maximum out of each piece of hardware because we don't want to pay for an extra piece. If hardware and installation cost nothing, or at least if other concerns are more pressing, we don't care so much about optimizing the output of each piece. This scenario would lead to the second perspective: that we are trying to get as much energy as possible from a given amount of **space**. This is relevant on a rooftop or any residential setting, where the guiding question is "can I fit the PV system that I need into the space that I have?"

From a commercial perspective, CPV needs to address one of these needs in order to be useful. Does it?

To answer this we need to look into the unique physical and economic constraints that apply to concentrator systems.

2.3.1 Concentrator Physics: Fundamental Limits of CPV

The thermodynamics of optical concentration have been dealt with rigorously since the 1970s [16]. The basis for this investigation is the principle of etendue conservation: *for any optical system, the product of the spatial extent (beam cross-section or width) and angular extent of the propagating light cannot be decreased.* The consequence of this law is that any concentrator—a system that reduces the spatial extent of light—must have a limited *acceptance angle* (θ_A), or maximum incidence angle which can be concentrated onto the target. An ideal thermodynamic concentrator, which concentrates light in two dimensions, is limited by the fundamental concentration limit

$$C = \left(\frac{1}{\sin(\theta_A)}\right)^2$$

In order to make multijunction cells economically feasible, CPV systems concentrate light by more than 500x, and some systems even reach 1000 times. In commercial systems sub-ideal concentrators based on Fresnel lenses are used for the sake of compactness; however, this reduces the acceptance angle even further below the thermodynamic limit. This is effective at minimizing cell-related costs, but it raises two additional challenge: a requirement of precise sun-tracking, and of mismatch between the light coming from the sun and the acceptance of the concentrator.

Sun tracking is essential for CPV: since the sun (apparently) moves across the sky, and high-concentration optics have an acceptance angle in the range of 1–2° [17], the system must undergo some movement or modification in order to maintain the sun in the acceptance cone of the optics. This is presently achieved by mechanical means: the entire system or module is rotated to maintain normal orientation towards the sun [18]. In some systems that use only low concentration and concentrate light in only one dimension, tracking along a single axis is sufficient; however, for the high-concentration systems that we are primarily discussing here, two axis tracking is used. Because the acceptance cone is small, this tracking must be quite precise, and this requires a heavy piece of machinery which adds a substantial expense. In addition to the cost a conventional tracker is much too bulky and heavy to be used in most residential settings, limit CPV to large-scale fields. This takes CPV out of the running for rooftop installations, where high efficiency would be most useful.

In addition to this, there is a complication in system design that comes from using two-axis trackers. In a field where many CPV modules with trackers are installed, modules tend to shade each other, blocking light from the modules behind them as they rotate. Because a module consists of many cells connected in series, blocking one or two cells may substantially reduce the output of the entire module. Therefore shading should be avoided as much as possible. For this reason trackers are spaced out widely in a commercial installation, meaning that only a small fraction of the land is used to collect light. Shading is avoided when the sun is low, but when it is high most of its light falls on the ground between modules and it lost. Thanks to this spacing, a typical CPV installation does not actually save much space, per unit output, compared with flat-plate, and depending on how widely spaced the modules are, may even require more space! [19]

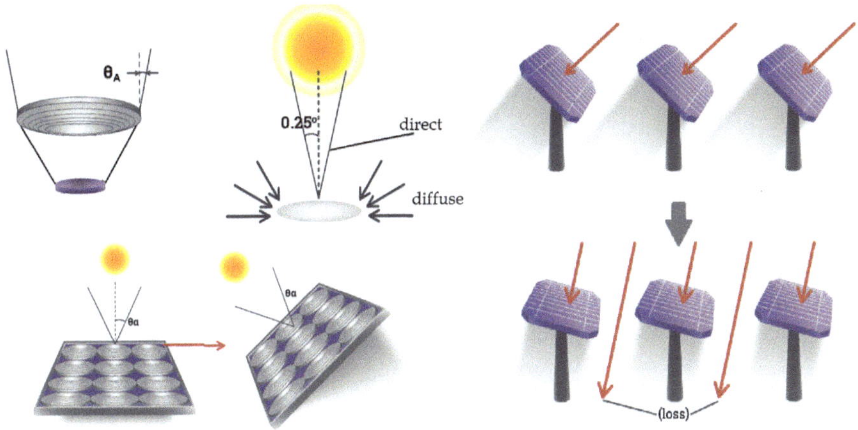

Fig. 2.4 Consequences of restricted acceptance angle in concentrators: since light incident outside the acceptance angle cannot be concentrated, diffuse light is lost and the concentrator must track the sun. To prevent panels from shading each other they are widely spaced, which means that most light incident at midday falls on the ground between the trackers and is lost

Finally, the acceptance angle of the concentrator is under many conditions not well matched to the angular profile of the incident sunlight. The "mismatch" between the incident light and the concentrator acceptance comes from the fact that much of the light coming from the sun is diffused by interactions with particles in the atmosphere. This diffuse light, falling outside of the concentrator's acceptance angle, cannot be concentrated and therefore is lost. So a concentrator actually has access to less light than a flat-plate collector under the same conditions Fig. 2.4.

2.3.2 CPV Economics

If CPV is not so useful as was hoped at reducing the size and space requirements of PV installations, can it reduce costs? That is, after all, what it was primarily designed for.

For an idea of this we can look to a recent NREL analysis of CPV module costs [20]. This study put cell-related costs in a 500X CPV module at roughly the same level, per Watt of output, as cell costs in a standard, non-concentrating module, but the high-concentration optics increase the module cost by nearly $0.20/W. The analysis put the minimum sustainable price of a CPV module at $0.77/W, far above the average market price of PV modules at the time ($0.57/W) [6]. And this is only part of the story. High-precision two-axis trackers impose an additional cost of greater than $0.30/W Fig. 2.5.

What does this mean for the economics of CPV?

The current cost given by NREL to install a large-scale PV power plant, including mounting, power conditioning equipment and installation costs, averages

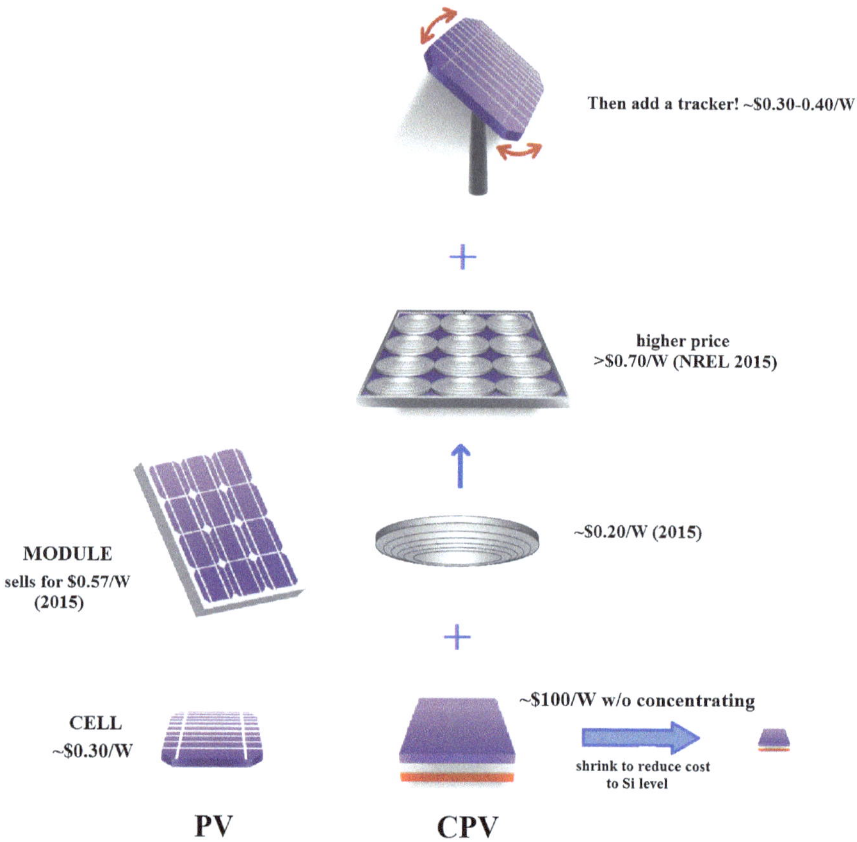

Fig. 2.5 Cost components of CPV versus flat-plate PV. Estimated module costs are somewhat higher for CPV due to the need for high-quality optical materials, and two-axis tracking adds a cost on top of the module that is not present for flat-plate

just under $1.50/W, and is coming down. [21] Pricing estimates for CPV systems are more difficult to come by due to the much smaller size of the market and lack of standardization. However we can get some idea just by naively considering the cost components described above—which represent the difference between CPV and a conventional system. Adding the cost of the tracker and the additional optics-related module cost represents a premium of about $0.50/W to use CPV rather than flat-plate—more than a 30% markup! The PV sector is very price sensitive, with aggressive bidding wars and resultant price reductions driving who gets business and who goes out of business. Add to this that flat-plate PV, as we have seen, is a mature technology riding a smooth learning curve to ever-lower costs as it scales [22]. CPV, as a cost-saving proposition, seems to be dead on arrival.

So what remains?

If we believed CPV were truly dead we would not be writing a book about it! The fact is that the problems we have outlined have been recognized by many researchers and a wide array of creative technological solutions have been proposed. These form the basis for taking bold new approaches to CPV to finally make concentrator technology into affordable PV products, to satisfy niches in the large and growing solar energy market that would otherwise be left unfilled. CPV, we believe, is far from dead; it has simply falle out of step with the realities and needs of today. With appropriate development, CPV can be an extremely versatile technology which opens new dimensions of design space for developing innovative products. The keys to realizing this potential lie in the pages of optics journal, unused patents and overlooked startups. Our purpose here is to collect these bits of creative technological invention and piece them together in new ways that better fit today's economic realities.

The technological components can be broken down into three: cells, optics and tracking. In each area there are exciting developments that, taken together, sketch out a framework for a new CPC paradigm. The three chapters that follow will highlight the themes and innovations in each area, and weave them into a bigger picture that will give context and ground for the main goal of this book—reimagining CPV.

References

1. Nelson, J. (2003). *The physics of solar cells* (Vol. 57). World Scientific.
2. Green, M. A. (1982). Solar cells: operating principles, technology, and system applications.
3. Shockley, W., & Queisser, H. J. (1961). Detailed balance limit of efficiency of p-n junction solar cells. *Journal of Applied Physics, 32,* 510–519.
4. Hirst, L. C., & Ekins-Daukes, N. J. (2011). Fundamental losses in solar cells. *Progress in Photovoltaics: Research and Applications, 19,* 286–293.
5. Kerr, M. J., Cuevas, A., & Campbell, P. (2003). Limiting efficiency of crystalline silicon solar cells due to Coulomb-enhanced Auger recombination. *Progress in Photovoltaics: Research and Applications, 11,* 97–104.
6. Barbose, G. L., & Darghouth, N. R. (2016). Tracking the Sun IX: The Installed Price of Residential and Non-residential Photovalic Systems in the United States (Washington: US Department of Energy). *available at emp. lbl. gov/publications/trackin.*
7. Green, M. A., et al. (2017). Solar cell efficiency tables (version 49). *Progress in Photovoltaics: Research and Applications, 25,* 3–13. doi:10.1002/pip.2855.
8. Green, M. A. (2001). Third generation photovoltaics: Ultra-high conversion efficiency at low cost. *Progress in Photovoltaics: Research and Applications, 9,* 123–135.
9. Chiu, P. et al. (2014). In *Photovoltaic Specialist Conference (PVSC), IEEE 40th.* 0011–0013 (IEEE).
10. Fraunhofer, I. S. E. (2014). New world record for solar cell efficiency at 46%. Press release.
11. Kayes, B. M. et al. (2011). In *Photovoltaic Specialists Conference (PVSC), 2011 37th IEEE.* 000004–000008 (IEEE).
12. Green, M. A., Emery, K., Hishikawa, Y., Warta, W., & Dunlop, E. D. (2015). Solar cell efficiency tables (Version 45). *Progress in Photovoltaics: Research and Applications, 23,* 1–9.
13. Bett, A., Dimroth, F., Stollwerck, G., & Sulima, O. (1999). III-V compounds for solar cell applications. *Applied Physics A, 69,* 119–129.

14. Bauhuis, G., Mulder, P., Haverkamp, E., Huijben, J., & Schermer, J. (2009). 26.1% thin-film GaAs solar cell using epitaxial lift-off. *Solar Energy Materials and Solar Cells, 93,* 1488–1491.
15. Murphy, T. (2011). *Don't be a PV Efficiency Snob* https://dothemath.ucsd.edu/2011/09/dont-be-a-pv-efficiency-snob/.
16. Winston, R., Miñano, J. C., & Benitez, P. G. (2005). *Nonimaging optics*. Academic Press.
17. Shanks, K., Senthilarasu, S., & Mallick, T. K. (2016). Optics for concentrating photovoltaics: Trends, limits and opportunities for materials and design. *Renewable and Sustainable Energy Reviews, 60,* 394–407.
18. Mousazadeh, H., et al. (2009). A review of principle and sun-tracking methods for maximizing solar systems output. *Renewable and Sustainable Energy Reviews, 13,* 1800–1818.
19. Ong, S., Campbell, C., Denholm, P., Margolis, R., & Heath, G. (2013).Land-use requirements for solar power plants in the United States. *Golden, CO: National Renewable Energy Laboratory* **140**.
20. Horowitz, K., Woodhouse, M., Lee, H., & Smestad, G. (2015). Bottom-Up Cost Analysis of a High Concentration PV Module; NREL (National Renewable Energy Laboratory). (NREL (National Renewable Energy Laboratory (NREL), Golden, CO (United States))).
21. Fu, R. et al. (2016). US Solar Photovoltaic System Cost Benchmark Q1 2016. (NREL/PR-6A20-66532. Golden, CO: National Renewable Energy Laboratory (NREL)).
22. Swanson, R. M. (2006). A vision for crystalline silicon photovoltaics. *Progress in Photovoltaics: Research and Applications, 14,* 443–453. doi:10.1002/pip.709.

Chapter 3
High-Efficiency Solar Cells

One idea that we will return to throughout this book is that CPV, as configured today, is an "overconstrained" system from an engineering standpoint. Too many design elements and environmental factors must be controlled too precisely to make a system that is simple and cheap enough to compete economically. The strongest physical constraints on CPV are the result of using high concentration. But the requirement of high concentration comes from the high cost of the multijunction cells that are used to give CPV its signature high efficiency. If we want to relax these constraints, we need to start by understanding why the cells are so expensive and how they can be made less so.

In the introduction we sketched out the contours of the problem: Silicon and other single-junction cells are limited by the mismatch between their narrow spectral response and the broad spectrum of sunlight; multijunction cells allow the spectral response to be broadened to more closely match the incident, but only certain materials, most commonly the III–Vs, can be used; these materials are expensive (GaAs costs something like 1000x more than solar-grade Si) [1], and therefore the cell are costly as well. The benefits to reducing these costs are clear, and there are two obvious approaches—either make III–V cells cheaper, or find other materials that serve the same function.

3.1 Making High-Quality Multijunction Solar Cells

Based on economics alone, the expensive III–Vs seem a strange choice to make solar cells, which need to be very cheap in order to generate electricity at a competitive cost. However the III–V materials have as unique combination of beautiful properties that have made them irresistible to both physicists and engineers, and in fact these properties make them well suited to multijunction cells in a way that no other known material family is [2].

H. Apostoleris et al., *Concentrating Photovoltaics (CPV): The Path Ahead*,
Green Energy and Technology, DOI 10.1007/978-3-319-62980-3_3

The III–V material par excellence is gallium arsenide. GaAs has been known for decades as the closest thing to an ideal optoelectronic material that we have. The record-efficiency single-junction solar cell was made from GaAs in 2011 [3]. What makes it so good, in particular as a solar cell material?

3.1.1 Band Gap and Absorption

Our discussion of band gaps and carrier generation in solar cells that we offered in the preceding chapter is missing an important piece of information: The difference between a direct and an indirect band gap. In our basic picture we only considered the change in energy that is required to promote an electron into the conduction band, but in some materials, such as Si, the electron must also change its *momentum* as well as its energy in order to make this jump. This means that the electron must both absorb energy from a photon and then quickly exchange momentum with the crystal lattice through a phonon. The probability of these two events coinciding is much lower than that of the single electron-photon interaction. Therefore the photon will in general penetrate much farther into an indirect band gap material before creating a carrier pair. As a result a Si solar cell must be on the order of 100 μm thick to allow full absorption of light (and real cells are even thicker than this due to manufacturing considerations).

GaAs, on the other hand, has a direct band gap. This is desirable for other electronic applications as well but for solar cells it means that absorption is much stronger than in Si. This allows cells to be made much thinner—a few μm thick in conventional designs, and even thinner in newer cells with optimized light management. This is the root of a longstanding argument for GaAs as a photovoltaic material—it is much more expensive than silicon, but much less is used to make a cell. This in principle makes it resonable to use a more expensive material to get more ideal behavior.

3.1.2 Photoluminescence and Recombination

Another contributor to the high quality of GaAs as a solar cell is what happens *after* a photon is absorbed. We noted in the second chapter that one of the phenomena that limits solar cell efficiency is the recombination of electron-hole pairs before they can be extracted into the external circuit. There are in fact many mechanisms by which recombination takes place. It can occur, for example, because of defects in the crystal structure of the semiconductor that modify the electronic states. In Si, the dominant recombination mechanism is Auger recombination whereby an electron and hole recombine through an interaction with another electron [4]. But in an ideal solar cell, the main mechanism is radiative recombination: An electron and hole recombine to emit a photon with energy close to the band gap. In good-quality

GaAs, nearly all recombination is radiative. This means that a high-quality solar cell can also be run in reverse as an LED (this was one of the guiding principles that led to the record-efficiency GaAs cell) [5]. Radiative recombination produces a photon that can be "recycled" through absorption by another electron. Other mechanisms cause the carrier energy to be lost as heat. So the best possible solar cell is one where all recombination is radiative [6, 7]. GaAs again comes closest to that mark.

The other III–Vs used for solar cells—InP, GaInP and GaAsP are the most commonly used [8, 9]—as the chemical 'cousins' of GaAs and share many of its desirable properties such as high photoluminescence. In fact the LED industry is largely based on III–V materials for this reason. When the group is considered together it adds a few more considerations that make it well-suited to multijunction solar cells in particular:

3.1.3 Band Gap Tunabilty

The entire point of a multijunction cell is to split light between devices with different band gaps. Until recently the III–Vs were the only group of materials whose band gaps could be varied over the entire range of the solar spectrum, and showed good photovoltaic performance over this range. All that is required to change the band gap is to vary the composition of the different elemental components, so the layers of the multijunction cell can be made by the same process just by varying the ratios of a few materials [10].

Lattice matching: In the next section we will look at a commercial and experimental CPV cell to illustrate the main constraints on cell design that are relevant to building good multijunctions. One of the most important is that the crystal structures of neighboring layers of the cell need to be well-aligned, if one is being deposited directly on top of the other—this is the so-called *lattice matching* condition. The III–Vs offers a number of compositions that have both near-optimum combinations of band gaps, and good lattice matching [10]. This makes the fabrication of high-quality multijunctions fairly straightforward. The lattice matching condition applies not only to the layers of the cells, but also to the substrate on which the entire cell is deposited. This proves to be a major factor in the cost of III–V cells.

3.2 Multijunction Cells—Design and Manufacture

Lattice-matched epitaxial growth: In order to more completely understand the current state of multijunction technology we can take a quick look at an industry-standard three-junction cell. While we focused in the description on III–Vs, the traditional multijunction is actually composed of two III–V layers of mid and wide band gap—GaInP and GaAs or GaInAs—on a Germanium lower cell [11]. This allows each layer to be deposited directly on the one below, since the

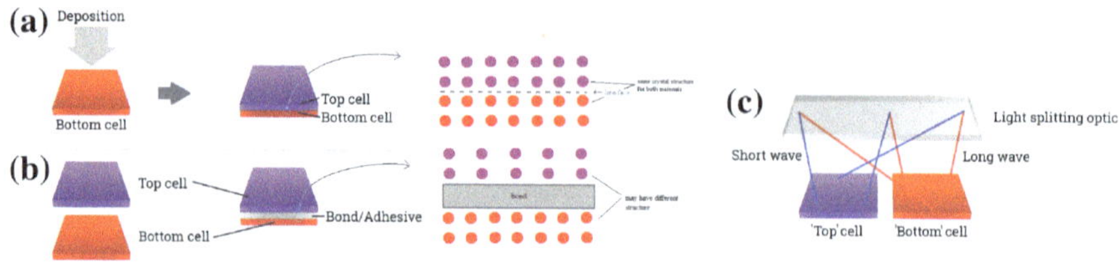

Fig. 3.1 Approaches to making multiple-junction cells (**a**) deposit a *top* (wide band gap) cell on *top* of the *bottom* cell directly, if crystal structures of two materials are well matched (e.g. GaAs on Ge) or if high-quality crystal is not needed (e.g. perovskite on Si); (**b**) fabricate two cells separately and bond or stick them together (e.g. GaInP on Si); (**c**) keep the two cells independent and use a light-splitting optic to divide light between them

three materials are lattice matched to each other. Historically the lattice matching to Ge is one of the reasons for the popularity of III–V, since Ge is a commonly used substrate for device fabrication [8] (Fig 3.1).

The application of multijunctions to both spaceflight and CPV has driven incremental improvements in performance over the last ten years or so.

3.2.1 *Lattice-mismatched Epitaxial Growth*

The approach of using materials with a slight lattice mismatch in multijunctions was successfully exploited in 2007 by Spectrolab to add a few points of efficiency by optimizing the band gap combination in a three-junction cell. It was these "metamorphic" cells that were the first to break 40% efficiency. This design used a graded-composition buffer layer to gradually relax the lattice spacing from the first to the second layer [11]. But as this approach can handle only small lattice mismatches, and building a cell of more than 3 junctions requires using materials with significant mismatch, other techniques were needed to further expand the design space and efficiency potential of multijunctions.

3.2.2 *Wafer Bonding*

Larger lattice mismatches can allow full exploitation of band gap engineering. The world-record multijunction solar cells (46% efficiency under 300x concentration) have dispensed with lattice-matching altogether through new developments in wafer-bonding technology [9, 12]. The cell is initially grown as two separate dual-junction "tandem" devices, on different substrates (GaAs and InP). After fabrication, the two cells are polished to create an extremely smooth surface, and the two cells are directly adhered. In this way large lattice mismatches are incorporated into a single cell without producing detrimental defects. The wafer bonding process is however quite intensive as a very smooth and clean surface is required for good bonding between the two cells [13, 14]. This may pose additional challenges in bringing these cells to low-cost mass production.

3.2.3 Mechanical Stacking

The simplest approach is simply to take two independent cells and stick them together! This is the approach of mechanical stacking. It differs from wafer bonding in that an adhesive or mount is used to connect the two cells—there is no bond at the atomic level. This is one of the oldest approaches to making multiple-junction devices [15, 16], and after some time in the wilderness interest is now growing again [17, 18].

Mechanical stacking has been used by researchers recently to investigate dual-or multijunction cells based on Silicon [18, 19] This is, to us, one of the more promising approaches since it "piggybacks" on the success of Si photovoltaics rather than trying to replace it with a new technology. Silicon is potentially useful in such applications since it has a fairly low band gap; in fact, it is nearly ideal as the bottom junction of a two-junction cell [20]. Mechanical stacking can allow the two cells to be made separately and then combined at the end of the process. Recently this approach has been used at NREL to demonstrate a nearly 30% efficient tandem of a GaInP top cell on a Si bottom cell [21].

3.2.4 Laterally-array Cells with Spectrum Splitting

Something of a challenge in fabrication of high-performance tandem cells is the potential for electrical, optical and damage related losses that arise from growing one cell on top of another, or attaching two separately-fabricated cells together. In a concentrating system, this can be circumvented by using a spectrum-splitting optic. The two cells can then be used in parallel without requiring stacking. Potentially, this can also provide a pathway to higher-efficiency multijunction systems using a larger number of cells.

3.3 Towards Ultrahigh Efficiencies: Feasibility of Many-Junction Systems

Optical spectrum splitting has been surveyed at length in the literature [22, 23] as it has been a hot topic with many researchers [24, 25], including the authors of this book [26], pursuing it as a promising approach to high-efficiency systems. In the past it has been presented as a way to avoid lattice matching as well as current-matching requirements in multijunction cells. Even so, to date there has been little progress on making a commercially viable spectrum-splitting PV system. This is largely because the strongest limitation on current mutijunction technology is the lack of *any* low-cost solar cell materials with the appropriate range of band gaps. More recently there has been some interest in making solar cells with lateral variation of the composition, so that the band gap varies from one edge of the cell to

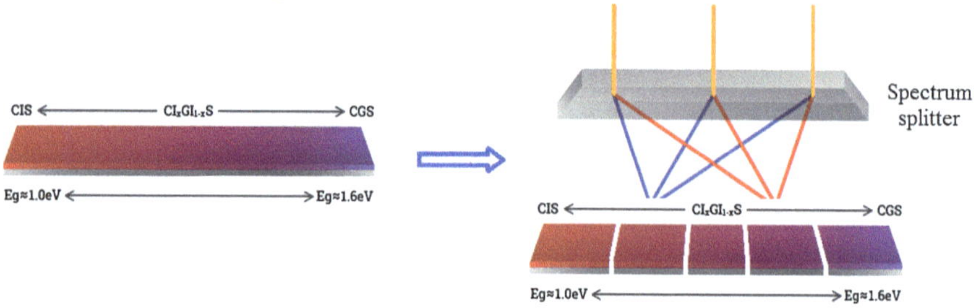

Fig. 3.2 Example of band-gap grading combined with spectrum splitting. CIGS is graded laterally from CIS (1.0 eV) to CGS (1.6 eV), then diced into arbitrarily many cells of different band gap and combined with a spectrum splitting optic

the other [27]. These could then be subdivided into arbitrarily many smaller cells, each with a different band gap, and combined with a spectrum splitting optic to produce a many-junction system at the cost of a single cell. One candidate material system that has been explored is the chalcogenides, which includes the popular flexible solar cell material CIGS ($CuInGaSe_2$). CIGS itself can be varied between 1.0 eV and 1.6 eV band gap (although with good photovoltaic performance over only a part of that range) [28, 29]; we use this as an example to illustrate in Fig. 3.2. However doing this over an appropriate range of band gaps has proven challenging, and what is more it is not entirely clear how much of a benefit this would provide. This is because of an important question that, we think, is not asked frequently enough: how many junction do we actually need?

3.4 How Many Junctions Do We Need?

At what point should we stop adding layers and focus on optimizing a smaller number of sub-devices, in order to get the best performance?

If we look at the record-efficiency cells we see an interesting trend. Silicon, the most-developed technology, has record cell efficiencies about 90% of its theoretical limit. The overall record for a single junction cell is Alta Devices' GaAs cell which has 28.8% efficiency, or 85% of the theoretical limit for a general single-junction cell [3]. But Fig. 3.3 shows what happens when we go to more junctions. The record dual-junction cell gains only 3% of efficiency (absolute) over the record single-junction, and while high-performing 3, 4 and 5-junction cells have all been demonstrated they are all quite close in efficiency—adding another junction can quickly become a game of diminishing returns [30, 31]. Partly this is the result of material constraints that cause real cells to deviate from the ideal band gap combinations which are assumed in the optimum efficiency calculations; [32] partly it is due to optical losses and reduction in material quality that come from adhering or growing a cell on top of another cells. In part it is just a fundamental feature of

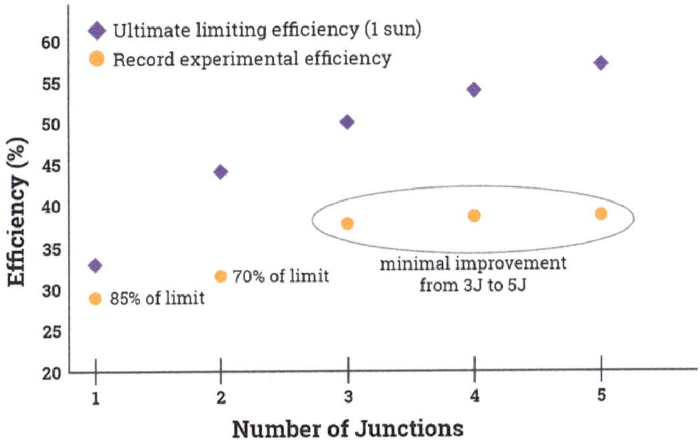

Fig. 3.3 Efficiency records from NREL's photovoltaic cell efficiency chart 2016 [34]; ultimate efficiencies from Marti and Araujo [33]

multijunctions: For each incremental junction that is added, the incremental gain in ultimate efficiency is less [33].

One possible conclusion of this is that we have pushed too aggressively towards many-junction systems without fully addressing sources of loss in the two-junction system. If a two-junction cell were developed with the same performance relative its limiting efficiency as our best single-junction cells (Silicon, with record efficiency at 90% of its physical limit), its efficiency would be near 40%. Therefore it is worth considering that perhaps if a single optimized dual-junction technology were adopted and developed at an industrial scale, high performance could be achieved with much lower cost and complexity.

But when the big picture is considered, how to put the two cells together may be the least troublesome problem. Of much greater concern is *which cells are we going to use*? Devices such as NREL's GaInP-on-Si are useful to demonstrate particular methods or principles, but they are of little use from a commercial standpoint since they rely, again, on a III–V top cell which may be a hundred times the cost of the bottom Si cell! [21, 35] This leads us into the big challenge of high-efficiency PV: No matter what clever ways we find to put cells together, it is not worth much if we do not have the right cells!

3.5 Multijunction Cell Materials: The Big Picture

For all the technologies and materials that have been explored, there are currently only three technologies that are economically viable for terrestrial use: Silicon, Cadmium Telluride and CIGS. All three of these are used for reasonably priced commercial modules (with CdTe holding about 10% of market share and CIGS a distant third, used mainly in niche applications where flexible cells are needed) [36]. But their band gaps range only between the 1.1 eV or Si and the 1.5 eV of CdTe. This is generally considered not a big enough difference to produce enough efficiency gain to make a two-junction system based on these technologies economically viable [37]. In order to make a high-performing dual-junction device, high-quality cells with wider band gaps are needed. The only material group that is used to make such cells commercially is the III–Vs.

The problem in high-efficiency cells, then, is the absence of the right material for wide-bandgap cells. The current technology is too expensive, which leads to the requirement of high concentration in systems that use them; cheaper materials have not yet proven adequate for commercial PV. So there are two directions to look: Can we make III–V cells much, much cheaper? And what new material developments are on the horizon that will offer new pathways?

3.5.1 *Cheaper III–V Cells*

The main difficulty of III–Vs, as we have seen, is their expense—per unit area, III–V cells cost on the order of 100x more than a crystaline Si or commercial thin-film solar cell. High-efficiency GaAs cells that represent the state of the art in high efficiency single junction PV provide a baseline case. To present a general picture, the baseline cost for III–V cells is upwards of \$10,000/m2—in the neighborhood of \$50/W at typical 1-sun efficiencies [38] (Figs 3.4 and 3.5).

Manufacturing III–V cells begins with a substrate wafer which is lattice matched to the material that the cell will be made of—Ge or GaAs are used for GaAs solar cells, and InP is popular for indium-containing compounds [9, 39]. A buffer layer of a different material is then deposited by vapor-phase epitaxy, followed by the layers of the cell. To make a thin film cell, an etching process is then used to remove the buffer layer, separating the cell from the substrate. This process is called "epitaxial liftoff" In principle, the wafer can be reused to grow another cell; however in practice it usually has sustained some damage that prevents high quality films from being grown on it a second time. Therefore the wafer is usually thrown away [39].

Since the substrate materials are very expensive, this is extremely wasteful! 80–90% of the cell cost is the cost of the substrate on which the cell is grown [39, 40]. If the process is modified and optimized so that the substrate wafer is not damaged during cell

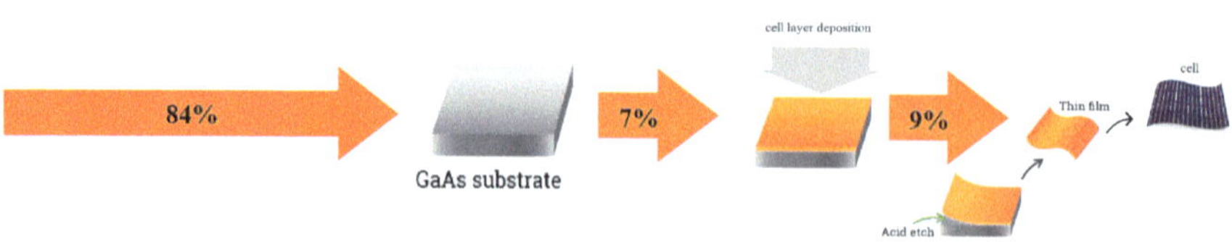

Fig. 3.4 III–V cell costs

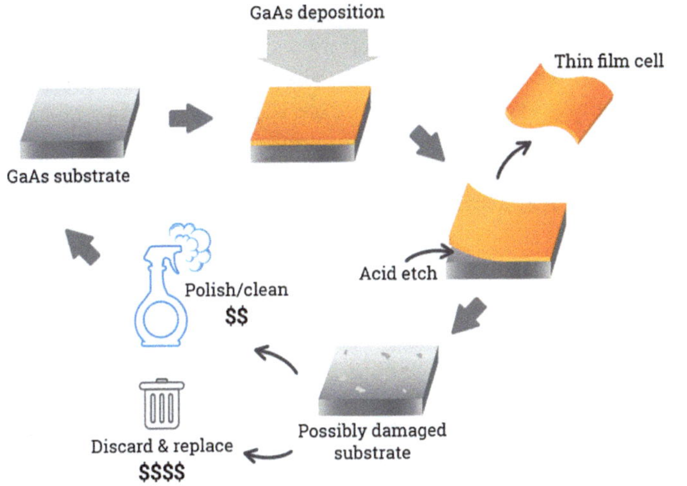

Fig. 3.5 Epitaxial liftoff and wafer recycling process: This is the general approach used by makers of large-area GaAs solar cells, e.g. Alta Devices

liftoff, then it can be reused to grow more cells. If this process can be repeated a large number of times, the cost contributed by the wafer can become very low [38].

Wafer reuse is becoming more common as manufacturers and researchers try to move III–Vs towards commercial viability. Since the cells themselves are very thin, repeated wafer reuse can remove nearly all material costs from the cell manufacturing process. But the process is still some way away from producing a large-area III–V cell at a price consumers will pay. NREL's analysis of substrate reuse methods concluded that even a fairly mature wafer reuse process costs somewhere between 10 and 30% the amount of simply buying new wafers. This is progress, but not the 100x reduction that would be needed to start putting GaAs or multijunction solar panels on people's rooftops!

Recent work [40, 41] has focused on new approaches to epitaxial liftoff and other thin-film production methods that may open new avenues to low-cost wafer recycling. But even in the event that it can be achieved, the manufacturing process itself is far more expensive for III–Vs than for Si or other commercial thin-film technologies. The last 10–15% of the cost is attributed to cell deposition process itself, which relies on expensive metal-organic vapor phase epitaxy (MOVPE). Reducing these costs to the point of competing with $0.50/W commercial panels would likely require massive economies of scale, which still might never catch up to the much more advanced Si learning curve.

Other methods of producing III–V cells that have the potential to reduce these costs have been explored but are at a less developed level. Recently NREL has begun testing experimental manufacturing methods that replace MOVPE with other, less capital-intensive properties. The proposed method also dispensed with the need for a lattice-matched substrate, possibly circumventing the need for wafer recycling altogether [42]. However the devices grown by these experimental processes have to date shown inferior performance to conventionally manufactured cells, so again a revolution in the cost of III–Vs does not seem to be imminent.

This does not necessarily mean that we are trapped in an all-or-nothing situation regarding high-efficiency cells. In the short term, it may still be feasible to use CPV, reducing the required cell area through concentration. The economics of this solution might become favorable if there were, say, an order-of-magnitude reduction in costs. This would not make III–Vs viable for 1-sun configurations, but by reducing the required concentration ratio from $\sim$500X to $\sim$50X, it could significantly reduce the constraints on optics and tracking, possibly opening a cost-effective pathway for high-efficiency CPV. But since any new technology would have to compete directly with Si, which has a huge head start in terms of development, even a solution that works on paper might not be able to get off the ground fast enough to catch up economically in the near future.

3.5.2 *Beyond III–V—Other Materials for Low-Cost Multijunction*

The second option is to look to other materials for high-efficiency solar cells. This approach has gotten a boost recently by the discovery, just in the last few years, of extremely good photovoltaic performance in perovskites. In this time, perovskites have begun to be seen as a rival of III–Vs in photovoltaic (and other optoelectronic) performance, with efficiencies climbing to 22% in only a few years of research [43].

The term “perovskite” can refer to a naturally occurring mineral, calcium titaniate, but the term also refers to a wide range of compositions that have the same crystal structure as the “original.” Perovskites used for solar cells are typically a mixed organic-inorganic material; for example, the most-developed composition and the one with the best performance to date is methylammonium lead iodide ($CH_3NH_3PbI_3$) have a number of appealing properties that may make them the solution to high efficiency at this particular point in the development of the PV industry [44]. With a wide variety of possible compositions, they are band gap tunable and can cover much of the wide-gap range that is needed to complement silicon. Critically, they can be deposited by extremely simple process—they are closer than anything we have to the dreamed-about “paint-on” solar cell. So not only do they avoid the high cost of III–Vs, they would likely be, once production is scaled up, even cheaper than silicon to produce! [45] There is no lattice-matching requirement, which allows them to be deposited on top of Si or even on top of a full module to create a dual-junction device at low cost.

If the breakneck progress of the last few years continues, the assumptions around high-efficiency PV could be turned upside down. But there are still significant problems with perovskite cells. The best perovskite absorbers (MAPbI3) are degraded very quickly by exposure to water [46]. In addition, there is some difficulty with diffusion of some components out of the cell and into the electrodes. These are sizable challenges to surmount if we expect to see perovskite solar cells on the market soon. But given the fast pace of development on this particular technology, and the absence of a good alternative, it is worth anticipating as a reasonable possibility in the near future.

This survey of high-efficiency cell technologies and developments has, we think been useful for providing context but in terms of our goal of reimagining CPV it has been a bit of a false start. The hope for making multijunction cells significantly cheaper using current technology seems fairly remote, and the most likely pathway seems now to be the perovskites, if they can be made reliable and durable enough to be implemented as a commercial product. This would potentially change the economics of high-efficiency PV radically, since a perovskite-based cell could easily cost less to manufacture than Si technology. This would effectively decouple high efficiency from concentration, since the most efficient cells—probably a perovskite-on-Si device as is under investigation by some of the biggest players in perovskite PV—would also be among the cheapest, and therefore no concentration would be necessary for their use. But if this scenario does not pan out there may still be ways to reinvent the way we use our current high-efficiency cells, even with only more modest cost reductions. We will begin to consider how this might be done by looking at the optics.

References

1. Woodhouse, M., & Goodrich, A. (2014). *Manufacturing cost analysis relevant to single-and dual-junction photovoltaic cells fabricated with III–Vs and III–Vs grown on Czochralski silicon (presentation)*. CO: National Renewable Energy Laboratory (NREL), Golden.
2. Bett, A., Dimroth, F., Stollwerck, G., & Sulima, O. (1999). III–V compounds for solar cell applications. *Applied Physics A, 69,* 119–129.
3. Kayes, B. M., et al. (2011). *Photovoltaic Specialists Conference (PVSC), 37th IEEE* (pp. 000004–000008). Piscataway: IEEE.
4. Yablonovitch, E., & Gmitter, T. (1986). Auger recombination in silicon at low carrier densities. *Applied Physics Letters, 49,* 587–589.
5. Yablonovitch, E., Miller, O. D., & Kurtz, S. R. (2012). *In Photovoltaic Specialists Conference (PVSC), 38th IEEE* (pp. 001556–001559). IEEE: Piscataway.
6. Shockley, W., & Queisser, H. J. (1961). Detailed balance limit of efficiency of p-n junction solar cells. *Journal of Applied Physics, 32,* 510–519.
7. Hirst, L. C., & Ekins-Daukes, N. J. (2011). Fundamental losses in solar cells. *Progress in Photovoltaics: Research and Applications, 19,* 286–293.
8. Cotal, H., et al. (2009). III–V multijunction solar cells for concentrating photovoltaics. *Energy & Environmental Science, 2,* 174–192.
9. Dimroth, F., et al. (2014). Wafer bonded four-junction GaInP/GaAs//GaInAsP/GaInAs concentrator solar cells with 44.7% efficiency. *Progress in Photovoltaics: Research and Applications, 22,* 277–282.
10. Vurgaftman, I., Meyer, J., & Ram-Mohan, L. (2001). Band parameters for III–V compound semiconductors and their alloys. *Journal of Applied Physics, 89,* 5815–5875.
11. King, R., et al. (2012). Solar cell generations over 40% efficiency. *Progress in Photovoltaics: Research and Applications, 20,* 801–815.

12. ISE", F. (2014).
13. Zahler, J. M. *et al. NCPV and Solar Program Review Meeting.*
14. Derendorf, K., et al. (2013). Fabrication of GaInP/GaAs//Si solar cells by surface activated direct wafer bonding. *IEEE Journal of Photovoltaics, 3,* 1423–1428.
15. Gee, J. M., & Virshup, G. F. (1988). *Photovoltaic Specialists Conference, Conference Record of the Twentieth IEEE* (pp. 754–758). Piscataway: IEEE.
16. Antypas, G. A., Bell, R. L., & Moon, R. L. (1982). Google Patents.
17. Lee, K.-H., et al. (2016). In *Photovoltaic Specialists Conference (PVSC), IEEE 43rd* (1957–1959). Piscataway: IEEE.
18. Essig, S. *et al.* Realization of GaInP/Si dual-junction solar cells with 29.8% 1-sun efficiency. *IEEE Journal of Photovoltaics, 6,* 1012–1019 (2016).
19 Essig, S., et al. (2016). In *Photovoltaic Specialists Conference (PVSC), 2016 IEEE 43rd.* (2040–2042). Piscataway: IEEE.
20 Vos, A. D. (1980). Detailed balance limit of the efficiency of tandem solar cells. *Journal of Physics. D. Applied Physics, 13,* 839.
21 Essig, S., et al. (2015). Progress towards a 30% efficient GaInP/Si tandem solar cell. *Energy Procedia, 77,* 464–469.
22 Mojiri, A., Taylor, R., Thomsen, E., & Rosengarten, G. (2013). Spectral beam splitting for efficient conversion of solar energy—A review. *Renewable and Sustainable Energy Reviews, 28,* 654–663.
23 Imenes, A., & Mills, D. (2004). Spectral beam splitting technology for increased conversion efficiency in solar concentrating systems: a review. *Solar Energy Materials and Solar Cells, 84,* 19–69.
24 Kosten, E. D., Warmann, E. C., Lloyd, J., & Atwater, H. A. *SPIE Solar Energy+Technology.* 882109-882109-882103 International Society for Optics and Photonics.
25 Keevers, M. J. *et al.* High Efficiency Spectrum Splitting Prototype Submodule Using Commercial CPV Cells. (2015).
26 Stefancich, M., et al. (2012). Single element spectral splitting solar concentrator for multiple cells CPV system. *Optics Express, 20,* 9004–9018.
27 Caselli, D., & Ning, C.-Z. (2015). Monolithically-integrated laterally-arrayed multiple bandgap solar cells for spectrum-splitting photovoltaic systems. *Progress in Quantum Electronics, 39,* 24–70.
28 Rampino, S., Bissoli, F., Gilioli, E., & Pattini, F. (2013). Growth of Cu (In, Ga) Se2 thin films by a novel single-stage route based on pulsed electron deposition. *Progress in Photovoltaics: Research and Applications, 21,* 588–594.
29 Gabor, A. M., et al. (1996). Band-gap engineering in Cu (In, Ga) Se2 thin films grown from (In, Ga) 2Se3 precursors. *Solar Energy Materials and Solar Cells, 41,* 247–260.
30 NREL Photovoltaic Efficiency Chart. (2016).
31 Green, M. A., et al. (2017). Solar cell efficiency tables (version 49). *Progress in Photovoltaics: Research and Applications, 25,* 3–13. doi:10.1002/pip.2855.
32 Kayes, B. M., Zhang, L., Twist, R., Ding, I.-K., & Higashi, G. S. (2014). Flexible thin-film tandem solar cells with > 30% efficiency. *IEEE Journal of Photovoltaics, 4,* 729–733.
33 Marti, A., & Araújo, G. L. (1996). Limiting efficiencies for photovoltaic energy conversion in multigap systems. *Solar Energy Materials and Solar Cells, 43,* 203–222.
34 NREL. (2016).
35 Bobela, D. C., Gedvilas, L., Woodhouse, M., Horowitz, K. A., & Basore, P. A. (2017). Economic competitiveness of III–V on silicon tandem one-sun photovoltaic solar modules in favorable future scenarios. *Progress in Photovoltaics: Research and Applications, 25,* 41–48.
36 Trube, J., Fischer, M., & Metz, A. (2016). Wiley-V Ch Verlag Gmbh Postfach. 101161, 69451 Weinheim: Germany.
37 Peters, I., Sofia, S., Mailoa, J., & Buonassisi, T. (2016). Techno-economic analysis of tandem photovoltaic systems. *RSC Advances, 6,* 66911–66923.

38 Horowitz, K., Woodhouse, M., Lee, H., & Smestad, G. (2015). Bottom-Up Cost Analysis of a High Concentration PV Module; NREL (National Renewable Energy Laboratory). (NREL (National Renewable Energy Laboratory (NREL), Golden: CO (United States).

39 Ward, J. S. (2016). Techno-economic analysis of three different substrate removal and reuse strategies for III–V solar cells. *Progress in Photovoltaics: Research and Applications, 24,* 1284–1292.

40 Lee, K., Zimmerman, J. D., Hughes, T. W., & Forrest, S. R. (2014). Non-Destructive Wafer Recycling for Low-Cost Thin-Film Flexible Optoelectronics. *Advanced Functional Materials, 24,* 4284–4291.

41 Kim, Y., et al. (2017). Remote epitaxy through graphene enables two-dimensional material-based layer transfer. *Nature, 544,* 340–343.

42 Zheng, M., et al. (2016). *III–Vs at scale: a PV manufacturing cost analysis of the thin film vapor–liquid–solid growth mode*. Progress in Photovoltaics: Research and Applications.

43 Green, M. A., Ho-Baillie, A., & Snaith, H. J. (2014). The emergence of perovskite solar cells. *Nat Photon, 8,* 506–514. doi:10.1038/nphoton.2014.134.

44 Chilvery, A. K., et al. (2015). Perovskites: Transforming photovoltaics, a mini-review. *PHOTOE, 5,* 057402–057402. doi:10.1117/1.JPE.5.057402.

45 Hwang, K., et al. (2015). Toward large scale roll-to-roll production of fully printed perovskite solar cells. *Advanced Materials, 27,* 1241–1247.

46 Wang, D., Wright, M., Elumalai, N. K., & Uddin, A. (2016). Stability of perovskite solar cells. *Solar Energy Materials and Solar Cells, 147,* 255–275.

Chapter 4
New Approaches to CPV Optics

4.1 Solar Concentration: Practical Optics and Physical Limits

We first discussed the constraints imposed by using concentrator optics in the first chapter, where we noted that a fundamental, thermodynamic limit to concentration exists which is given by [1]:

$$C = \left(\frac{n}{\sin(\theta_A)}\right)^2$$

where n is the refractive index and θ_A is the acceptance angle defined in Chap. 1. The only exception to this is the luminescent concentrator [2], which has been investigated for some time but is subject to other limitations, and will not be explored here. The concentration limit is derived from the principle of etendue conservation, which has been written about in depth by previous authors [3, 4], so we give only a short summary.

Etendue is a property that measures, roughly, how "spread out" light is. For a beam that is propagating with a cross-sectional area A_1 and a divergence given by the half-angle θ_1, the etendue is given by:

$$E = A_1 \sin^2(\theta_1)$$

Thermodynamically, etendue measures the number of quantum states in the beam and in this way is connected to the optical entropy.. As such it is inherently non-decreasing. This means that the thermodynamically best optical device will keep the etendue of the light constant as it propagates:

$$A_1 \sin^2(\theta_1) = A_2 \sin^2(\theta_2)$$

H. Apostoleris et al., *Concentrating Photovoltaics (CPV): The Path Ahead*, Green Energy and Technology, DOI 10.1007/978-3-319-62980-3_4

This says that there is a tradeoff between the spatial and angular extent of the beam—in order to concentrate the light to a smaller area, it is necessary to expand its divergence. The concentration can be written as A_1/A_2. This means that the concentration ratio of an ideal optical device can be related to the angular divergence of the input and output beams:

$$C = \frac{\sin^2(\theta_2)}{\sin^2(\theta_1)}$$

The maximum possible concentration is reached when the output beam is diffused fully so that it fills a hemisphere, with $\theta_2 = 90°$. Then the numerator is 1 and the concentration ratio is

$$C = \frac{1}{\sin^2(\theta_1)}$$

where θ_1 is the same as the acceptance angle θ_A since it represents the angular range from which the device can accept light and concentrate it onto the exit surface. But if this light then passes into a medium with a refractive index n, its angular divergence will be reduced due to refraction by a factor of n^2. Then the light can be further concentrated by this factor, leading to the concentration ratio limit above.

This is a typical starting point for discussions of CPV optics, but it is only part of the story. Some optical devices can come close to this ideal concentration—the most well-known example of this is the compound parabolic concentrator, which was the first optical device to be optimized for energy collection rather than for forming images and so is called a nonimaging optic [5]. Nonimaging optics typically have a major drawback for applications to solar collectors: their length is proportional to the concentration ratio, so in high-concentration systems they are completely impractical to use [6]. Therefore they can be useful for low-concentration applications [7], but high-concentration CPV systems must use modifications of older, imaging optics to collect light. These have a much smaller acceptance angle than the ideal limit [6, 8]. Figure 4.1 illustrates this by plotting the thermodynamic limits of concentration in air and in dielectric media, along with the acceptance angles and concentration factors of real CPV systems.

The main types of optics used in CPV systems are also the oldest—lenses and mirrors. Today, the Fresnel lens is the most popular optic for CPV. Fresnel lenses typical of CPV systems tend to have an acceptance angle below 0.5° [8]. To correct this, commercial CPV systems use a small secondary optic near the cell which tends to direct the focused rays onto the cell. However these still bring the acceptance angle only up to about half of the ideal limit, while also introducing more complexity to the system.

This illustrates one of the difficulties that arises in CPV optics design. A number of factors—system complexity and cost, compactness, optical throughput, acceptance angle, irradiance profile—must be simultaneously optimized. These, however, tend to have tradeoffs with each other. This is illustrated in these examples of the

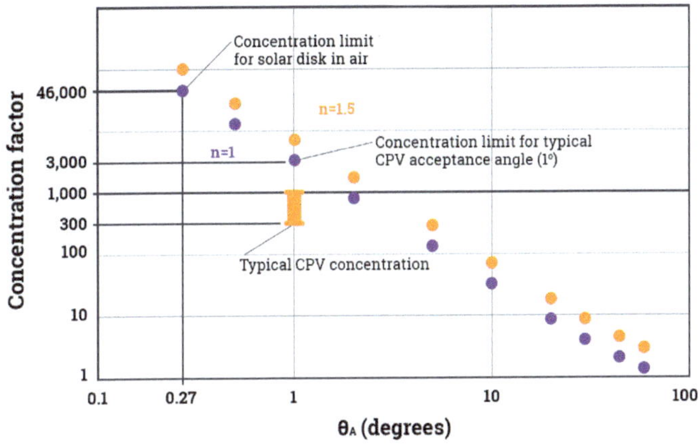

Fig. 4.1 Thermodynamic concentration limit as a function of acceptance angle. We point out the ultimate concentration limit for the solar disk (divergence angle 0.27°) and the typical acceptance angle of CPV optics (~1°). The range of concentrations achieved by commercial CPV systems is several times below the limit for these acceptance angles

Fig. 4.2 *Tradeoffs in CPV optics*

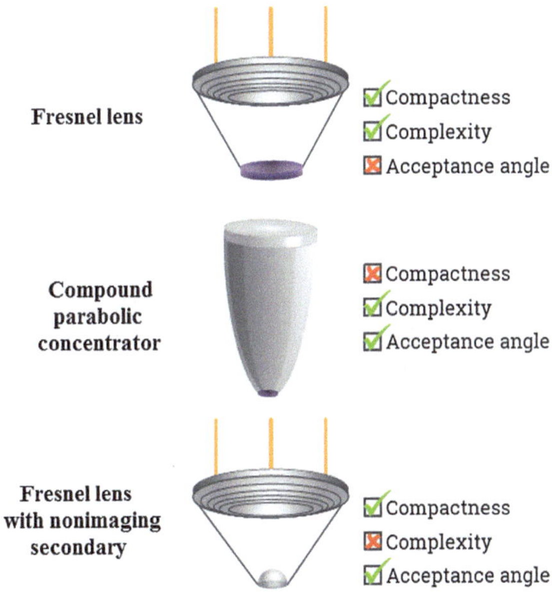

Fresnel lens (a simple system and compact, but with a small acceptance angle), the CPC (optimum acceptance angle, but not compact except at low concentrations), and the compound optic consisting of an imaging primary optic with non-imaging secondary and sometimes tertiary optics (more compact, acceptance angle better than the nonimaging optic alone, but with more complexity) (Fig. 4.2).

4.2 Constraints of Small Acceptance Angles

In the previous chapter we discussed the mismatch between the light spectrum incident on a solar cell and the light spectrum that is converted by the cell—a solar cell is inherently suited to convert light with a narrow spectral distribution, but sunlight has a broad distribution. We also noted briefly that a solar concentrator suffers from an analogous mismatch: a concentrator, due to the principle described above, admits light from a small acceptance angle, which in principle should be sufficient since the angular extent of the sun in the sky is quite small (0.27°). However this only accounts for a part of the sunlight that reaches the PV system. There are actually two different ways in which the angular profile solar resource is mismatched with the angular response of the concentrator. We can think of one of these as being a "dynamic mismatch," and the other an "instantaneous mismatch".

Because of the small size of the solar disk, at any moment, sunlight coming directly from this disk can be captured by a very high-concentration device (the maximum thermodynamic concentration of sunlight is 46,000x). But the sun, of course, does not stand still in the sky [9], which means that a device with a small acceptance angle must constantly reorient in order to keep the sun inside of its field of view. This is the dynamic mismatch: although the sun is *instantaneously* confined to a small angular space, the path it traces *over time* covers a very large angle and therefore its light cannot be efficiently collected by a *stationary* high concentration device. This is what leads to the requirement of sun tracking in CPV systems. The next chapter will be devoted to a more detailed survey of sun tracking; for now we note, as we did in the introduction, that sun tracking machinery suitable for CPV adds a substantial cost to the system and is a major contributor both to the high price of CPV and its impracticality for the settings in which it would be the most useful.

We did not consider in the introduction this second mismatch. While less blatant, it is critical for any practical assessment of a CPV system. The sunlight which comes directly from the solar disk is only part of the radiation that hits the earth's surface. A large part of the sun's light is diffused by particulates in the atmosphere. The fraction of the sun's light that is diffused, on a sunny day, may vary from less than 10% in a place with clear skies, to greater than 30% in a location with a high density of atmospheric particulates. Diffuse light, again from thermodynamic principles, cannot be concentrated since it is already highly disordered. In a conventional high-concentration CPV system, this diffuse component is therefore lost completely. In addition to this, there is a third component to sunlight that can lead to overestimation of system performance. The direct component of sunlight is measured by a device that has a small acceptance angle (but still larger than most concentrators), to filter out diffuse light. However, the acceptance angles of these measurement devices are still typically many times larger than the size of the solar disk itself, usually several degrees. This means that some amount of light that is measured as "direct" does not actually come from the solar disk itself, but from this space of a few degrees around it, which we call the "circumsolar region [10]."

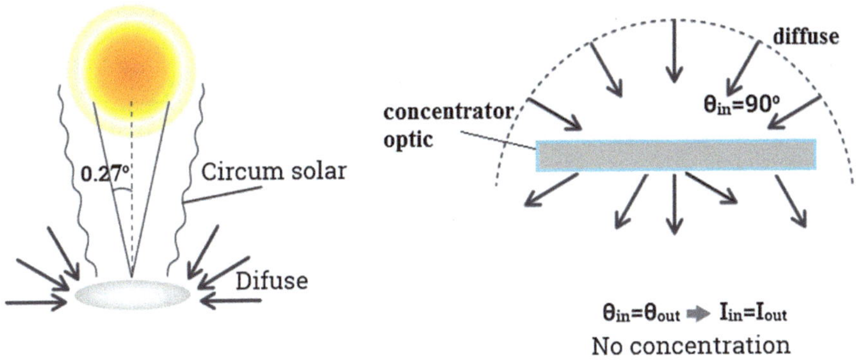

Fig. 4.3 The solar resource consists of a direct component (coming from the solar disk), a circumsolar component (weakly scattered light coming from the area around the solar disk, and the diffuse component (strongly scattered and coming from all directions). The diffuse component cannot be concentrated because its etendue is already maximized

While this distinction is invisible to standard measurement equipment, it is not invisible to a solar concentrator, which may have a smaller acceptance angle than the measurement device. Therefore a device with a very small acceptance angle may also fail to capture some of this circumsolar radiation, further reducing light collection (Fig. 4.3).

There is no way to solve this 'instantaneous mismatch,' whereby diffuse and, sometimes, circumsolar light are not captured by the concentrator. It is in this way (if we may stretch the analogy) more similar to the 'spectral mismatch' of the previous chapter. It may occur to the reader that it could, then be addressed in an analogous way, but we will return to that point later in this chapter.

Recognizing the loss of the diffuse component indicates a few important considerations that need to be made for CPV that are not relevant for flat-plate PV systems. One is that environmental factors need to be considered more aggressively for CPV, since air quality and cloud cover, which strongly impact the direct-diffuse ratio, have a much stronger impact on CPV. In the same way, it is also more important to prevent sand or dust from accumulating on CPV panels, since these also serve to diffuse light and prevent it from being concentrated onto the cells. Since standard test conditions may undersestimate the diffuse component of light, a project developer also needs to consider that the true efficiency of a CPV system will be less than its official rating, by an amount that is determined by these environmental factors.

It also gives a caveat to designers of modules. Some manufacturers have attempted to reduce costs by pushing towards higher and higher concentrations [11]. Our assessment is that this must be done very carefully, not only because increasing the concentration will tend to reduce the tolerance to tracking errors, but also because the reduction of the acceptance angle may actually cause less light to be collected by cutting into the circumsolar component. Therefore if higher concentrations are sought, they should be combined with developments in optics, which some companies seem to be engaged in, which use new designs to move the acceptance angle closer to ideality.

In the end, however, we think the bigger point is that the inherent restrictions imposed by high-concentration optics are what leads to CPV's impracticality and uncompetitveness. A true reimagining of CPV will depend on lowering the required concentration ratios. There are scenarios in which this could be done: if high-efficiency cells suddenly become much cheaper; or if manufacturers see some benefit in using cheap single-junction cells in combination with concentration, in order to get some other benefit. What happens when we relax the economic constraints that lead to the requirement of high concentration, and what new design possibilities are opened up?

4.3 Relaxing Optical Requirements with New Design Approaches

Current CPV technology uses concentrations of $\sim$500x to offset the cost of multijunction cells. The previous chapter described possible futures where (1) III–V cells remain the preferred option for multijunction efficiencies, and their cost is reduced by roughly a factor of 10 but not the factor of 100 that would be needed to compete with Si directly at 1 sun; or (2) entirely new, low cost materials such as perovskites become preferred for high-efficiency PV, potentially removing the cost factor entirely. In the first case, CPV systems might be similar to current HCPV systems, with small cells, point-focus optics and dual axis trackers. However since the cells could be 10x larger, we have C = 50X rather than C = 500X, and the engineering requirements are relaxed relative to the HCPV case. In the second, still lower concentrations could economically be used. But *why* one would use concentration with such cell is a question that is still to be answered.

We want to emphasize here the possibilities for novel thinking in concentration that are possible when our motivation is factors other than strict economics. What if the goal of concentration is not to mitigate the cost of an expensive cell, but to reduce the use of an already-cheap cell technology for environmental or long-term strategic reasons?

For example a company might want to move to a low-concentration module design in order to expand its module production capacity by several times without investing in new cell manufacturing capacity. From this perspective, 2X concentration could be twice as good as no concentration at all (as opposed to the HCPV case where it has no value whatsoever).

In addition to this straightforward low-concentration approach there is additional space to consider in the design of PV optics. This space encompasses the various ways of splitting light between different system components to maximize the energy usage. This can include, as many research have pursued, spectrally splitting light between PV cells of different band gaps; [12] or, using both PV and thermal absorbers for simultaneous electricity and heat production; [13, 14] or, the use of concentrators to separate diffuse from direct light either in a conventional module [15] or for building-integrated PV [16]. In addition, more fundamental thinking about the physics of concentration can point to new directions for research that may yield better optical devices that are freed from the tradeoffs that constrain conventional concentrator design. The space to explore is broad and we will therefore take the approach of a free-wheeling review touching on many different proposed optical designs for new PV concentrators, and some simple back-of-the-envelope economic analysis to give a picture of the potential of different approaches to cost-effectively utilize the solar resource.

We will start off by considering how much "space" there is to play in from an economic perspective. That is, when all manufacturing and materials costs are considered, what kinds of concentrator designs are feasible from a system-cost standpoint. A good benchmark for assessing alternative optical configurations is to require that any modification is price-neutral. Therefore the minimum cost of the optics manufacture informs the minimum allowable concentration ratio—the total cell costs will need to be reduced by roughly as much as it costs to add the concentrating optics.

The key to designing effective concentrators is to tailor the optics to the conditions where they are being deployed. When we can consider low concentrations as potentially useful, it becomes possible to design a concentrator for almost any situation.

For example, a system being deployed on a roof will have different assumptions regarding sun tracking than one being deployed in a field.

Likewise, a concentrator system designed for a cloudy or hazy area will need to be much different than one intended to be used under clear skies.

Of course, neither of these is a situation that is normally considered as appropriate for concentrators! But they can be possible with some creative thinking.

We will consider here the case of a traditionally acceptable ground-based system, and then move on to stationary concentrators and systems designed to handle diffuse light later in this chapter and the next.

First, a ground-based solar field in a clear, sunny region. In recent years an industry consensus seemed to be emerging to be that in such a setting, single axis trackers that follow the daily motion of the sun are a winning proposition, due to declining tracker costs and increased light collection. This means that a line-focus concentrator module can be substituted for a conventional module without increasing the BOS. However this has changed with continued decline in module prices. At least one company, several years ago, introduced a large-scale low concentration system which claims a lower LCOE than conventional PV (again, whether this claim is still true remains to be seen). Producing a low-concentration module with the form factor of a standard panel would have the benefit of being compatible with a generic tracker, which in such a case would still have value.

We briefly explore the economic feasibility of these low-concentration options. Our metric will be a cost per watt of *on-site* output, based on set environmental conditions and component costs per unit area.

For cost considerations of such a scheme, we look again to NREL's CPV cost analysis [17]. The cost of CPV optics is largely dictated by materials, which in the NREL paper are optical silicones molded into Fresnel lenses. In that analysis, the cost of the optics and cover glass, which are integrated into a single component, is given as 17 c/W at 30% efficiency, translating to $\sim$\$50/m^2. Current cost of conventional PV modules per square meter is, assume \$0.50/W at 15% efficiency, about \$75/m^2. This leaves very little space to reduce costs by adding any other elements (tracking, optics, cooling systems), since these will typically cost the same or more as the solar cells themselves. This shows, as many companies have discovered, getting cost reductions through concentration is a very tricky proposition. But if the inclusion of optics is seen is improving functionality, rather than just a cost-cutting strategy, then perhaps some useful approaches can still be found.

4.4 Light Splitting: Getting More Out of the Resource

Seeking these useful approaches, we now return again to the issue of mismatches between the solar resource and the response of the solar collector.

One solution that has been extensively explored, and which we have touched on in the previous chapter, is the splitting of solar radiation between physically independent cells of different band gaps. This presents a possible alternative path to ultrahigh efficiencies (>50%) as opposed to the stacked multijunction cells. At present, efficiency records stand at 39% for unconcentrated light [18] and 46% for concentrated light [19, 20]. All three demonstrations achieving these records used systems of four junctions, with either optical spectrum splitting or stacking of independently fabricated cells. The ultimate limits for photovoltaics conversion are about 68 and 86% for unconcentrated and highly concentrated light respectively, if more junctions can be added [21]. However this many-junction scenario is unlikely to be feasible for stacked cells. To grow such cells monolithically, a huge number of high-quality layers would need to be deposited on top of each other. Fabricating the cells separately and stacking them would entail optical losses due to the multiple interfaces between independent cells. The most promising scenario for many-junction cells is most likely to develop methodologies to grow cells with lateral band gap grading, as described in the previous chapter. Such cells could then be divided into arbitrarily many independently contacted subcells to give eight, ten or even more-junction systems of laterally arrayed cells. These systems could then give high efficiency, if used in conjunction with spectrum splitting optics.

A wide array of spectrum-splitting designs have been presented in the literature, involving filters [22, 23], geometric refractive [12] and reflective optics [24], and holographic optics to split light [25]. As several reviews have also been written [26, 27] we see no need to perform an exhaustive survey but rather will focus on what would be needed to make a commercially viable product involving spectrum splitting.

The most important considerations, of course, need to be cost and ease of use if the product is to be attractive. Here the problem with spectrum splitting is the same as that of CPV optics in general in that there is a tradeoff between compactness, complexity and acceptance angle, where improving one tends to harm at least one of these other desirable qualities. For an example we can consider the examples of a geometrical optics based spectrum splitter such as the one we have demonstrated in several previous works [12, 28], and the filter-based systems that are more popular in the current literature [23, 29]. The prismatic optic is dispersive and therefore requires only a single element regardless of the number of cells used; in contrast, the filter-based system must add an additional element for each 'band' that the spectrum is subdivided into, as we show in Fig. 4.4. This means that the complexity

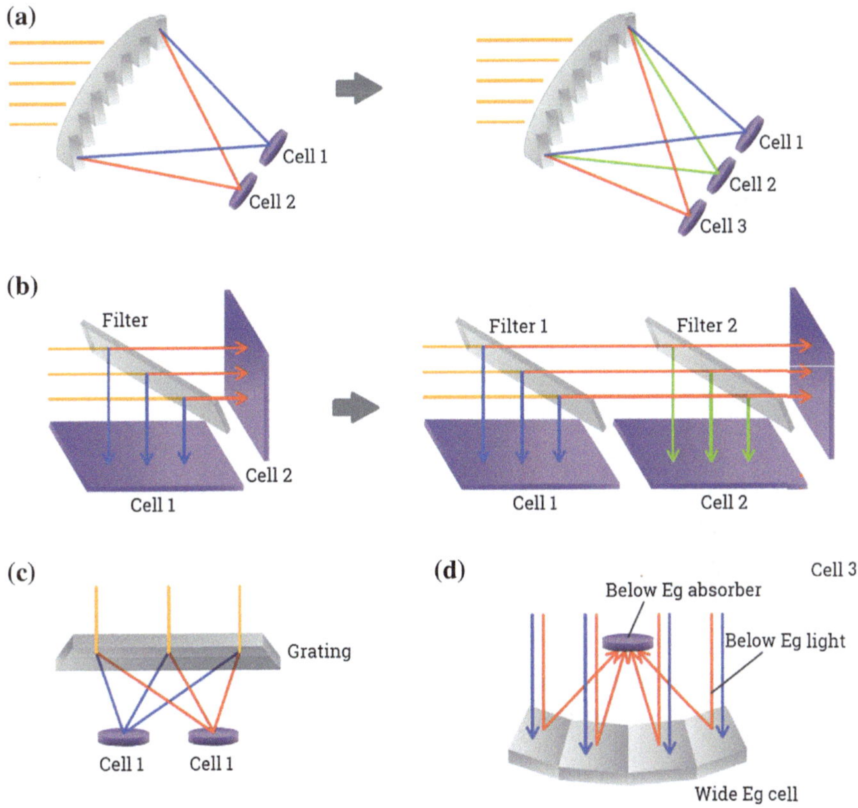

Fig. 4.4 **a** Options for spectral separation of light: Dispersive optics can illuminate many cells with a single optical device but may have a long focal length (not compact); **b** systems using filters may be more compact but increase in complexity as more cells are used. **c** Diffractive grating systems may offer compactness and simplicity simultaneously. **d** Solar cells may themselves be used as filters, which absorb above-Eg photons and reflect or transmit below-Eg photons to a second PV cell or thermal absorber

of the system and consequently the cost and opportunities for optical loss, increase with the number of cells used. The dispersive system does not have this scaling of the complexity, but is often less compact (our device, for example, has a focal length of nearly one foot). Even other optics designs that are more compact still suffer from the common feature of most CPV systems, of being much thicker than a standard PV module. The solution to this problem may be found in wavelength-scale engineering of materials for a desired photonic effect, rather that the use of macroscopic elements. One of these possibilities which has already been demonstrated on several occasions is a holographic element used as a spectrum splitter, where diffraction is the mechanism by which light is split. These can be produced in large areas and function with a short distance between the optic and the receiver. Light-trapping systems can also be used for spectrum splitting as has been proposed by Goetzberger et al. [30].

While the most discussed approach to light splitting, the multi-PV configuration is not the only possibility. Another strategy that has received some significant attention is dividing the system into a PV and a thermal component. This may involve using the PV cell as a filter and the sub-gap light for thermal [31]. Research on this front has started and stalled many times, but with the growing acceptance of solar energy it is now seeing a revival in, for example, domestic rooftop PV + water heating systems [14, 32].

A third approach does not involve any spectral separation. This returns to the point made earlier that the mismatch between diffused sunlight and the narrow angular response of a concentrator is analogous to the mismatch between the narrow spectral response of a solar cell and the broad spectrum of sunlight. An analogous solution may therefore be possible: to use a separate device to collect the diffuse light that is 'mismatched' to the concentrator, in the same way that a second cell can be used to collect below-bandgap light that is 'mismatched' to the spectral response of the solar cell.

A few proposals have been put forward to make use of this diffuse light. One group has proposed a system that exploits the low cost of silicon PV, making a CPV module using Fresnel lenses, high-efficiency concentrator cells and large-area silicon cells covering the receiver surface of the module [15]. Direct light is concentrated onto the high-efficiency cells; diffuse light is not concentrated but is transmitted through the optics and strikes the Si cells. In this way the loss of the diffuse is mostly compensated and a major failing of CPV is avoided. Another design that was first considered some time ago is to use a transmissive concentrator in the facade of a building. The direct light is concentrated onto the cells while the diffuse light passes through, illuminating the building. The facade is used for power generation without fully obstructing the window [16, 33].

These concepts can all be seen as new approaches to light splitting that go beyond split-spectrum photovoltaics, and may allow more complete use of the solar resource.

4.5 Angle Restriction and Concentration—an Emerging Concept

We would like in this last section to briefly discuss a concept that has been proposed theoretically but only recently achieved experimental demonstration. At the beginning, we discussed the thermodynamics of light concentration; the key point of this is the relationship between angular restriction and concentration. We can in fact go one step further and assert that, in the case where etendue is conserved, angular restriction and concentration are fundamentally equivalent. This was described thoroughly, in pieces, by various authors [25, 30, 34]. It can be shown in different ways that if light with a small divergence is admitted into a structure that restricts emission to occur only within an angle θ, and then randomized, and the reflections and angle restriction are perfect, the light trapped inside the structure will be $1/\sin^2(\theta)$ times more intense than the light incident from outside. By comparing to the ideal concentrator that we saw at the beginning of this chapter, we see that these two devices, although one is designed to move light from a large area to a small area, and the other to restrict the emission angle of randomized light, lead to the same optical result: light of the same intensity and same angular profile. This leads to the canonical statement that: concentration implies (acceptance) angle restriction, and (emission) angle restriction implies concentration! These devices may have the advantage of simultaneously optimizing many of the parameters described at the beginning of this chapter. Materials still lag behind what would be needed to realize solar concentrators based on this principle. However, with further materials development, angle-restrictor concentrators could provide a simultaneous solution to the problem of many competing parameters in PV optics design.

Our "big picture" summary with regard to CPV optics is that there is design space to explore, but not within the confines of high-concentration systems with high-cost cells. The developments in cell technology that were discussed in the previous chapter will play a large role in determining whether there is a place for such optical innovations in high-efficiency PV. There is another possibility, that concentrator optics will become useful in conjunction with single-junction, low cost PV as well, for example as new possibilities related to spectrum-splitting (or light-splitting generally) are explored. A number of these possibilities would likely be geared towards rooftop or building-integrated systems.

In order to include concentration in such settings, we will need to also rethink our approach to sun tracking.

References

1. Winston, R., Miñano, J. C., & Benitez, P. G. (2005). *Nonimaging optics*. (Academic Press, 2005).
2. Markman, B., Ranade, R., & Giebink, N. (2012). Nonimaging optics in luminescent solar concentration. *Optics Express, 20,* A622–A629.

3. Markvart, T. (2008). The thermodynamics of optical étendue. *Journal of Optics A: Pure and Applied Optics, 10,* 015008.
4. Winston, R., & Gordon, J. M. (2005). Planar concentrators near the étendue limit. *Optics Letters, 30,* 2617–2619.
5. Winston, R. (1974). Principles of solar concentrators of a novel design. *Solar Energy, 16,* 89–95.
6. Rabl, A. (1976). Comparison of solar concentrators. *Solar Energy, 18,* 93–111.
7. Antonini, A., et al. (2009). Rondine® PV concentrators: Field results and developments. *Progress in Photovoltaics: Research and Applications, 17,* 451–459.
8. Shanks, K., Senthilarasu, S., & Mallick, T. K. (2016). Optics for concentrating photovoltaics: Trends, limits and opportunities for materials and design. *Renewable and Sustainable Energy Reviews, 60,* 394–407.
9. Galilei, G. (1744). *Dialogo dove ne i congressi di quattro giornate si discorre sopra i due massimi sistemi del mondo Tolemaico, e Copernicano... in questa impressione migliorato ed accpeciuti sopra l'exemplare dell'autore stesso.* (Stamperia del Seminario, 1744).
10. Buie, D., Monger, A., & Dey, C. (2003). Sunshape distributions for terrestrial solar simulations. *Solar Energy, 74,* 113–122.
11. Stevenson, R. (2015). Turning To Ultra-High Concentrations To Increase The Competitiveness Of CPV in Compound Semiconductor.
12. Stefancich, M., et al. (2012). Single element spectral splitting solar concentrator for multiple cells CPV system. *Optics Express, 20,* 9004–9018.
13. Kraemer, D., et al. (2011). High-performance flat-panel solar thermoelectric generators with high thermal concentration. *Nature Materials, 10,* 532–538.
14. Crisostomo, F., et al. (2015). Spectral splitting strategy and optical model for the development of a concentrating hybrid PV/T collector. *Applied Energy, 141,* 238–246.
15. Yamada, N., & Hirai, D. (2016). *Maximization of conversion efficiency based on global normal irradiance using hybrid concentrator photovoltaic architecture.* Progress in Photovoltaics: Research and Applications.
16. Chemisana, D., Rosell, J., Riverola, A., & Lamnatou, C. (2016). Experimental performance of a Fresnel-transmission PVT concentrator for building-façade integration. *Renewable Energy, 85,* 564–572.
17. Horowitz, K., Woodhouse, M., Lee, H., & Smestad, G. (2015). Bottom-Up Cost Analysis of a High Concentration PV Module; NREL (National Renewable Energy Laboratory). (NREL (National Renewable Energy Laboratory (NREL), Golden, CO (United States)), 2015).
18. Chiu, P. et al. (2014). In *Photovoltaic Specialist Conference (PVSC), 2014 IEEE 40th.* 0011–0013 (IEEE).
19. Fraunhofer, I. S. E. (2014). New world record for solar cell efficiency at 46%. Press Release.
20. Dimroth, F., et al. (2014). Wafer bonded four-junction GaInP/GaAs//GaInAsP/GaInAs concentrator solar cells with 44.7% efficiency. *Progress in Photovoltaics: Research and Applications, 22,* 277–282.
21. Marti, A., & Araújo, G. L. (1996). Limiting efficiencies for photovoltaic energy conversion in multigap systems. *Solar Energy Materials and Solar Cells, 43,* 203–222.
22. Kang, D. W., Takiguchi, Y., Sichanugrist, P., & Konagai, M. (2016). InGaP//GaAs//c-Si 3-junction solar cells employing spectrum-splitting system. *Progress in Photovoltaics: Research and Applications, 24,* 1016–1023. doi:10.1002/pip.2753.
23. Kosten, E. D., Warmann, E. C., Lloyd, J., & Atwater, H. A. (2013). In *SPIE Solar Energy + Technology.* 882109-882109-882103 (International Society for Optics and Photonics).
24. Antonini, A. et al. In *Proc. 5th World Conference on Photovoltaic Energy Conversion.* 6–10.
25. Escarra, M. D., Darbe, S., Warmann, E. C., & Atwater, H. In *Photovoltaic Specialists Conference (PVSC), 2013 IEEE 39th.* 1852–1855 (IEEE).
26. Mojiri, A., Taylor, R., Thomsen, E., & Rosengarten, G. (2013). Spectral beam splitting for efficient conversion of solar energy—A review. *Renewable and Sustainable Energy Reviews, 28,* 654–663.

27. Imenes, A., & Mills, D. (2004). Spectral beam splitting technology for increased conversion efficiency in solar concentrating systems: a review. *Solar Energy Materials and Solar Cells, 84,* 19–69.
28. Carlo, M., Matteo, C., & Marco, S. (2015). Point-focus spectral splitting solar concentrator for multiple cells concentrating photovoltaic system. *Journal of Optics, 17,* 105901.
29. Keevers, M. J. et al. (2015). High Efficiency Spectrum Splitting Prototype Submodule Using Commercial CPV Cells.
30. Goetzberger, A., Goldschmidt, J., Peters, M., & Löper, P. (2008). Light trapping, a new approach to spectrum splitting. *Solar Energy Materials and Solar Cells, 92,* 1570–1578.
31. Abdelhamid, M., et al. (2016). Novel double-stage high-concentrated solar hybrid photovoltaic/thermal (PV/T) collector with nonimaging optics and GaAs solar cells reflector. *Applied Energy, 182,* 68–79.
32. Zheng, C., Li, Q., Rosengarten, G., Hawkes, E., & Taylor, R. A. (2014). In *Optics for Solar Energy.* RTu3B. 2 (Optical Society of America).
33. Dyson, A. H., Rh, P., Michael, S., & Jensen, K. (2007). Integrated concentrating (IC) Solar façade system.
34. Kosten, E. D., Atwater, J. H., Parsons, J., Polman, A., & Atwater, H. A. (2013). Highly efficient GaAs solar cells by limiting light emission angle. *Light: Science & Applications 2,* e45.

Chapter 5
Tracking Integration for Rooftop CPV

The last unique part of the CPV system that we have not discussed yet, and the part that most directly constrains CPV, is the sun tracker. The need for sun tracking is a direct consequence of the optical principles discussed in the last chapter: since a concentrator necessarily has a restricted acceptance angle, and the sun's apparent position in the sky varies over the course of the day, and with the seasons, tracking is the only way to keep a system operating for more than, in most cases, a few minutes per day.

We recall from the previous chapter:

The higher the concentrating power, the smaller the acceptance angle. The maximum acceptance angle is given by $\sin(\theta_a) = 1/C$ for a line-focus optic, by $\sin^2(\theta_a) = 1/C$ for a point-focus optic [1, 2]. This however can be enhanced by a factor of n or n^2 for an optic where the cell is immersed in a medium of refractive index n.

Only some non-imaging optics come close to this maximum acceptance angle, and lens-based optics preferred in most commercial CPV systems fall far short, usually with θ_a in the range of 1–2° for concentrations ~500X [3, 4].

A major component of CPV, which is also used in PV more broadly, is the mechanical system that is used to track the sun's motion, maintaining the solar disk inside the acceptance cone of the system. It is our opinion that the tracker is the single most important factor that has *directly* limited CPV growth. Here we will discuss tracking technologies, their costs, benefits and suitability for use with different types of systems, and consider how these costs, economic and otherwise, can be reduced or removed (Fig. 5.1).

5.1 Light Collection by Tilted Panels

The sun-tracker is essential to CPV thanks to its inherently limited acceptance angle. However, tracking has an impact on light collection for any solar device, not just a concentrator. In Fig. 5.2 we illustrate this by considering two solar panels of

H. Apostoleris et al., *Concentrating Photovoltaics (CPV): The Path Ahead*,
Green Energy and Technology, DOI 10.1007/978-3-319-62980-3_5

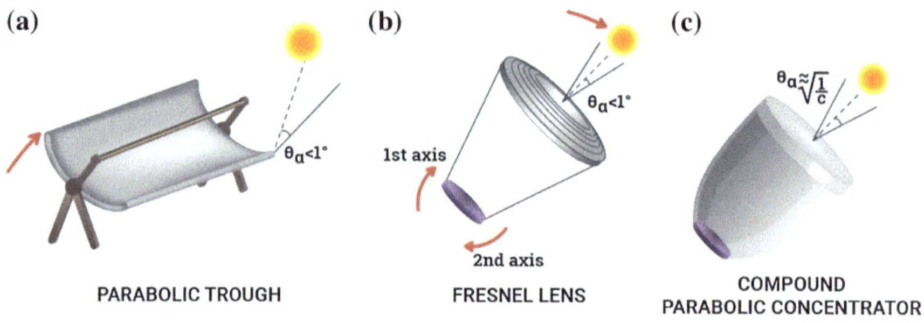

Fig. 5.1 Acceptance angle limits and tracking requirements of different concentrators

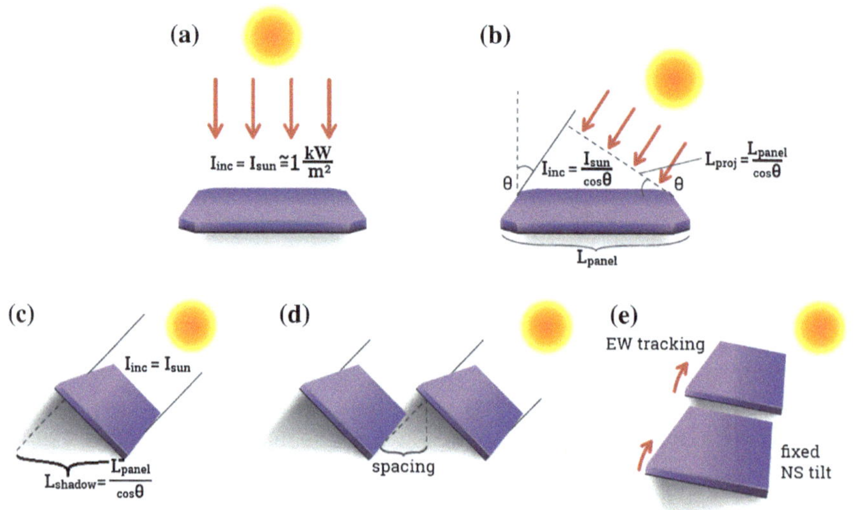

Fig. 5.2 Projection loss and tracking in flat-receiver systems. **a**, **b** A solar panel illuminated at an oblique angle will receive a light intensity that is reduced by the cosine of the incidence angle relative to one that is illuminated head-on. **c** If the panel is tilted towards the sun, this loss can be avoided, but at the expense of **d** casting a shadow that requires modules to be spaced out sufficiently to avoid shading. **e** light collection per panel can be optimized by tilting towards the sun in the north–south direction, then tracking the daily motion in the east–west direction

equal size but different orientation. The elevation of the sun is 45° above the horizon. One of the panels lies flat on the ground; the other is tilted by 45° to point directly at the sun. Now if we imagine "riding" the beam as it falls to the ground, or looking at the earth from the perspective of the sun, the apparent size of the two solar panels will be different. Precisely, the solar panel that is tilted towards the sun will seem to be its full size, as we are looking at it straight-on. The solar panel that is flat on the ground will appear smaller, by the cosine of the angle that we are viewing it from. Mathematically, we are seeing the projection of the panel's area onto our viewing plane. This is such an everyday experience that we usually ignore it, but it is a key understanding for sun tracking. Since light incident at an angle "sees" a compressed panel area, the panel "sees" a diluted or lower intensity light beam.

This is the reason why stationary solar panels are usually tilted, as closely as possible, to point towards the sun in the north-south direction. In a location like the northern US or Southern Europe (latitude 45°), the difference in noontime energy collection between a horizontal panel and one with a tilt equal to the latitude is a factor of about cos(45°) = $\sqrt{2}$. However this factor will be modified by the local conditions, including the amount of diffuse radiation from atmospheric scattering, and the albedo (reflectance) of the ground which produces additional diffuse reflected light. Both of these components will illuminate a tilted panel and do not follow the cosine behavior displayed by the direct beam [5].

The tilted panel also exhibits a kind of projection loss—this is inevitable whenever sunlight is not normally incident on the ground. This loss comes in the very familiar form of a shadow. We know from experience that shadows become longer as the sun gets lower in the sky; to be precise, the length of a shadow is $1/\cos(\theta_{inc})$ times the length of the panel, as we show in Fig. 5.2c. This means that, in an array, tilted panels must have a space between them equal to the shadow length if they are to avoid shading each other [6].

5.2 The Sun's "Motion"

The preceding discussion gave some preliminaries but we have not yet considered exactly what it is that needs to be tracked. The first thing that needs to be taken into account is how exactly the sun "moves" through the sky.

We are familiar from ordinary life with the basic movements of the sun, which consist of a daily and a yearly motion from the rotation and orbital periods, respectively, of the earth. During the day, the sun travels from east to west, reaching its highest point above the horizon at midday or solar noon. However, the sun is never directly overhead except in the tropics. In the northern hemisphere, the sun is always in the southern direction. The midday height of the sun above the horizon, we know, varies with the seasons, reaching its highest point on the summer solstice, which not coincidentally is the longest day of the year. These are daily-life observations of the solar behavior; for a more quantitative picture it is useful to define a coordinate system for the sky.

From our perspective on earth the sky appears as a hemispheric shell, which we view from the "inside." Therefore it is natural to use a spherical coordinate system defined by an azimuth angle, and an elevation angle (Fig. 5.3). The azimuth measures the angle along the horizon—north is an azimuth of 0, and east is azimuth 90°. The elevation is the angular distance above the horizon. So, for example, on the equinox at a latitude of 43° north, the sun's motion will be defined by the points: sunrise at azimuth 90° (east), elevation 0; midday at azimuth 180°,

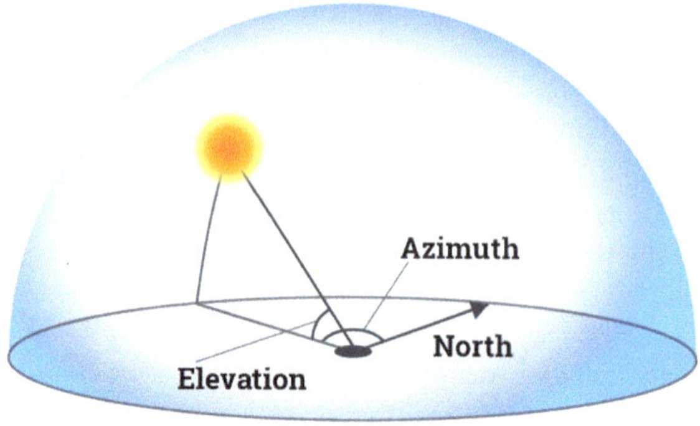

Fig. 5.3 Azimuth and elevation and the solar trajectory

elevation 43°; and sunset at azimuth 270°, elevation 0. Between the summer and winter solstices, the midday solar elevation changes by ±23.5° from the equinox position (latitude angle).

5.3 Sun Tracking on a Single Axis

In the first section we discussed only how to orient a stationary panel to optimize light collection. This is straightforward, if a limited acceptance angle does not need to be considered, since the sun, from summer to winter, varies by only ±23.5° from the latitude angle, and stays in the same half of the sky (north or south, depending on which hemisphere it is in) for the full year (except in the tropics). The next step is to account in the same way for the daily motion of the sun. There is no way to do this in a static configuration—the sun covers a full ±90° from east to west over the course of a day. To account for this we need to use a single-axis tracker. The panel rotates along the daily axis so that it is normal in this direction to the sun, eliminating projection loss along this axis.

A single axis tracker may be "flat"—in which case it suffers from projection loss in the north-south direction—or have a fixed tilt in the north-south direction to minimize this loss while tracking in the east-west axis. In the latter case, daily tracking can provide an additional gain of 20–35% light collection depending on conditions, relative to a fixed-tilt system without daily tracking. As a result some installers advocate using trackers for ground-mounted systems, regardless of whether concentration is used or not [7].

Now that we have looked at these preliminaries we can begin to consider concentrators where the additional requirement of the tracker is that the sun be maintained within a restricted acceptance angle. This is in general more stringent than simply optimizing light collection by a flat plate: while a flat receiver can tolerate moderate misalignment without a significant drop in light collection, a concentrator may entirely cease operation when the sun drifts outside of the acceptance cone.

The simplest case for tracking is a line focus concentrator. An example of these are the original solar concentrators, the parabolic trough which uses a dish to focus light onto a tubular receiver. These can employ daily tracking, usually with a flat north-south orientation. Alternatively, the troughs can be oriented east to west so that the sun moves along the length of the trough, and the north-south orientation is adjusted seasonally. The same options are available for a low-concentration CPV module that uses line-focus optics. Daily tracking, with a fixed north-south tilt, while unwieldy, is better for optimizing light collection per panel, since it minimizes projection loss. There is some argument over whether this is very important which we will discuss.

Because it allows for improved light collection per module, some non-concentrating PV installations use single-axis tracking. This can be useful when space constraints are not an issue, and instead the optimum configuration is one that uses as little hardware (modules and trackers) as possible. In addition, the resulting energy generation profile becomes more uniform over the course of the day, which may have value from the perspective of better transmission grid utilization and power generation profile optimization (Fig. 5.4).

The final step, which relates directly to the high-concentrations used in most CPV models, is tracking for point-focus optics. Since these concentrate in two dimensions, the acceptance is restricted along both axes and a two-axis tracker is needed. Most of these consist of a pedestal with a strong joint supporting the weight of the CPV module; the joint rotates horizontally to track the azimuth, and tilts to track the elevation. If the acceptance angle is large, the tracking need not be extremely precise. However, since most CPV modules have acceptance angles of $\sim 1°$ [8], the trackers used for these systems tend to be very heavy to provide mechanical strength, which allows them to maintain an accurate orientation in the face of confounding factors.

Adding trackers to any PV system introduces additional problems related to shading. As the panels rotate to track the sun, they will cast shadows on whatever is behind them. The lower the solar elevation, the longer the shadow. We can think of this shading in the same terms as projections loss as we showed previously, the length of the shadow is inversely proportional to the cosine of the elevation angle. The need to account for shading is in a broader sense the consequence of a basic

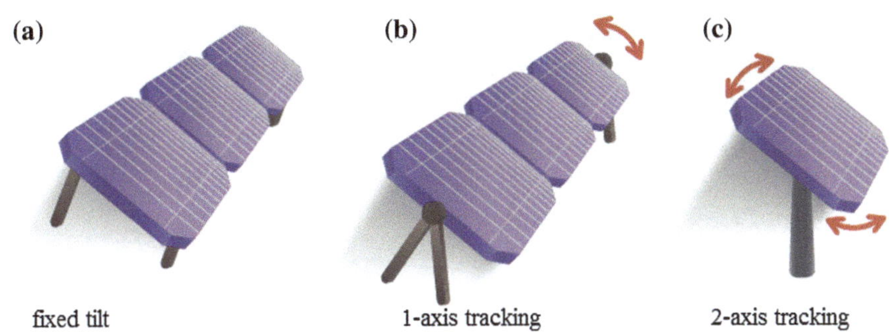

Fig. 5.4 Single axis and dual axis tracking options for flat panels and concentrators

physical reality: while we can, by tracking, increase the amount of light falling on a flat receiver, thereby increasing the system's capacity factor, there is no way to increase the amount of light that falls on a given area of horizontal ground. By tracking we can reduce the number of modules needed to collect the light, but we can never collect more light from a solar field of a given physical size than we would by simply covering the ground with solar panels pointed directly at zenith. There is simply no more light to collect. So the *per module* increase in output must be paid for by an equal or greater *per area* decrease in the number of modules installed.

Shading can be a significant problem especially if panels are internally wired in series—in this case, a small amount of shading will kill the output of the entire module section covered by a bypass diode, since current will be limited through the shaded cells. For this reason, trackers are widely spaced, especially dual-axis CPV trackers, which may cast long shadows in any direction due to their wide range of motion. The practical result of this is that tracking, either for CPV or non-CPV, can actually be a less efficient use of land than a fixed-tilt system, since trackers will be widely spaced to avoid shading at low solar elevations. In other words, tracking PV or CPV installations are designed for the most dilute light conditions that they will encounter, which means that much of the light that falls on the field during the full day will not be collected at all.

5.4 Physical Considerations of Sun Tracking 2: Tracking Errors

The optical considerations that have been discussed so far constitute an ideal scenario where we assume perfect tracking—that it is a simple matter to keep the concentrator module pointed at the sun at all times. In fact, achieving adequately precise tracking is not entirely straightforward. CPV modules are large and heavy and may warp or misalign under real conditions, possibly deforming permanently if not well supported. Considering the small acceptance angles that are standard, even small misalignments can severely degrade performance. In addition to their own weight, the module must withstand wind loading in order to maintain proper orientation. In order to resist these stresses, very heavy machinery is used for trackers, so that the pedestal and drive equipment forms the great majority of the mass of a HCPV system. This is the root of the high cost of dual-axis trackers, which have proven a substantial drag on the economics of CPV.

5.5 Sun Tracking Economics

The preceding discussions of light collection in various tracking systems provide a baseline for talking about the economics of sun tracking. One recent trend is the growth of interest from mainstream PV installers in using single-trackers for ground-based installations, solely for the increase in per-module light collection. Recent industry sources [9] put the cost of single-axis trackers for flat-plate systems at ~10c/W. Assuming an increase of 20% power output per module, and a system cost upwards of $1/W for ground-based systems, of which about half is the cost of modules, this suggests an advantage to using daily tracking in such systems. However, this is dependent on the cost of modules and racking materials; if the cost of modules and fixed-tilt installation is very low, even an inexpensive tracker will not be economically favorable.

For high-precision dual axis trackers, the economics are different. For these, the cost is considered to be roughly 35 c/W [10] and the light collection is only marginally better than a well-oriented single-axis tracker. However, these are required for the operation of CPV, so a large cost premium is being paid to enable the use of the CPV module, which itself has a comparable cost-per-watt to lower efficiency flat-plate modules. This is a key factor that prevents CPV costs from being reduced enough to beat flat-plate.

The cost of the tracker results largely from the fact that it is, necessarily, essentially a piece of heavy industrial machinery. In order to reduce these costs, it will be necessary to either rethink the concentrators and cells used as has been discussed in the previous chapters, to allow single-axis trackers to be used, or fully reinvent the way in which tracking in done to enable lower-cost solutions.

5.6 Concentrators for the Rooftop

One consequence that jumps out of the preceding discussion is the consideration of where concentrators will be the most useful. In ground-based installations, high-efficiency CPV has limited value for a few reasons. The setting in which CPV is potentially useful is in a high-DNI environment—with clear skies and little cloud cover. This will usually be in a sparsely populated, desert environment—consider North Africa or the southwestern US. Here, land is a cheap resource which does not contribute much to the cost of solar electricity. Therefore the efficiency of the hardware is not very important—what matters is the cost per unit power output. What is more, today's CPV does not reduce land use by nearly as much as the module efficiency gain would suggest, as we have hinted at previously, because of the wide spacing that is put between the trackers to avoid shading [11]. As we have seen, it is difficult to beat flat-plate silicon on cost, with current technology.

The place where high efficiency, in general, is most useful, and where it may be worthwhile to pay a premium, is a setting where we do not have much space, and need to squeeze every bit of power out of the space we have—inside a city, perhaps, or certainly on the rooftop. Here the first challenge is to make sure that the high efficiency of the PV actually translates to more efficient use of space. Current technology poses a problem in this regard due to the tracker spacing issue referenced previously.

Furthermore, as has been noticed by researchers for some time, a sun tracker is not a rooftop-friendly piece of equipment.

The way to make full use of high efficiency PV is to remove the tracker entirely. One clear way to do this is, of course, to use cells that are cheap enough to be deployed in flat-plate modules without any concentration! The pursuit of this—a cheap high-efficiency solar cell—is what motivated the first chapter of this book. But while there have been some promising developments, there still does not seem to be a clear path to bring that technology to market in the near future. Then, if we still want to develop high-efficiency products for the rooftop, we will need to ask how we can downsize CPV.

The way to make CPV fit on the roof is to integrate the tracking apparatus into the module itself. Such a tracking-integrated design would allow the module to be mounted directly onto a fixed rack or a rooftop. The light collection would be the same as a non-tracking panel and the output power, if the internal tracking mechanism is capable of tracking over a full range, would be higher thanks to the high-efficiency cells (Fig. 5.5).

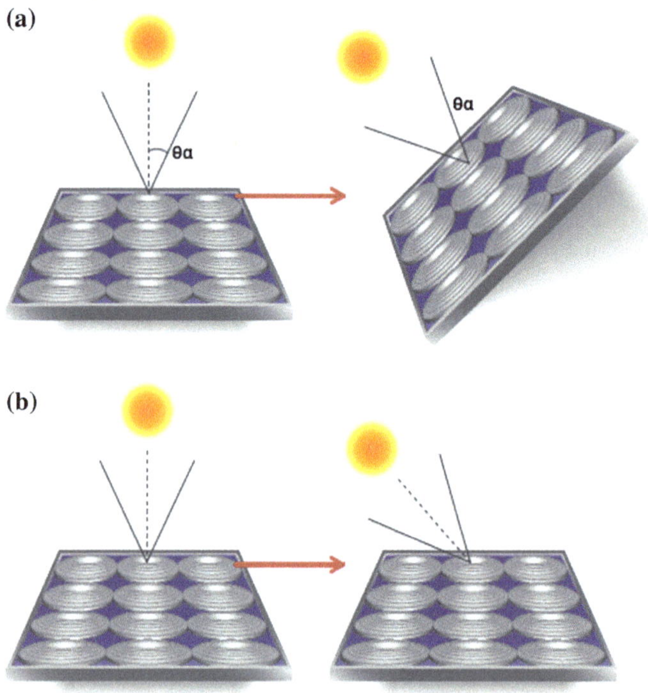

Fig. 5.5 Tracking integration: **a** In a standard CPV module, the narrow acceptance cone requires that the module be tilted to follow the sun in its daily and seasonal motion. **b** In a tracking-integrated module, the internal optics adjust to so that the acceptance cone shifts while the module itself stands still

We have previously reviewed the scientific literature and tested designs on integrated sun tracking for CPV [12]. Here we will go in detail into a few sample systems, to illustrate the considerations that must be made in order to achieve high-performance tracking integration, and what further developments will still be needed.

5.6.1 *Optical Principles of Tracking Integrated Solar Concentrators*

In order to fully understand the performance of tracked versus tracking integrated concentrators, and how this impacts their usefulness on a rooftop, we consider again the discussion from earlier in this chapter.

In terms of light collection, the tracking system in limited, as we have seen, by the need to avoid shading at low solar elevations. At the lowest elevation, this system collects the full incident intensity. However, at higher solar elevations, the wide spacing of the panels means that more and more light will fall on the ground between the panels and not be collected. The collected intensity is therefore limited to the intensity incident at the lowest elevation throughout the entire day! In the tracking integrated system, panels can be installed horizontally, or flush with whatever surface they are installed on (such as a roof). We have shown already that if the goal is to optimize light collection per unit land area, this is the best configuration since there is no situation where an incoming light ray will not intercept a solar panel (Fig. 5.6).

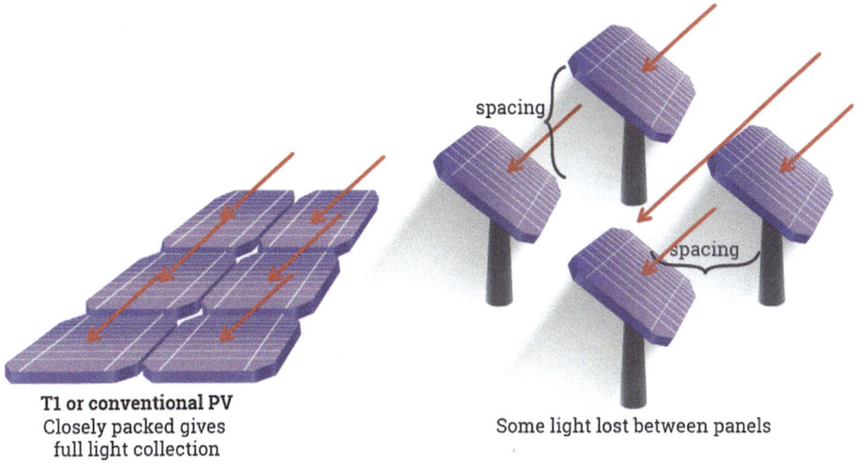

Fig. 5.6 Light collection in racking integrated versus tracked concentrators

In order to realize this it is necessary to have an internal tracking mechanism that can cover a wide angular range. As we have recently done an exhaustive survey we will only cover a few representative systems that illustrate the principles involved.

5.6.1.1 Design 1: Integrated Rotating Troughs

The first design that we will present here is one that we have recently presented and that we consider [13] one of the most straightforward integrated tracking scheme [13]. It is a line-focus system consists of a series of mini-troughs, perhaps a few centimeters in width, each one reflecting incident light onto a solar cell. In this way it mimics the well-established parabolic trough solar thermal systems, and some early CPV systems which used the same reflector-based design [14].

The tracking mechanism for this design is simply a miniaturization of a large installation of conventional trackers. One or more stepper motors is integrated into the module and connected via linkages to the troughs, which are free to rotate around an axis (Fig. 5.7a). Using an onboard controller following a simple tracking algorithm and using a small fraction of the power generated by the module, the motor rotates the troughs in unison to point at the sun. If a full ±90° tracking range is available, the light collection of this system will be equal whether the tracking mechanism is used to provide daily tracking or seasonal adjustments.

In order to achieve the maximum light collection, the tracking integrated system needs to have one significant difference from the ground based tracker array: the internally-tracking optics must be closely packed together, rather than being spaced out to avoid shading. This is the only way to have full collection of the available light at any solar elevation. This means that a tracking integrated module that is based on internal rotation of the optics will, except at normal light incidence, always have some shading of each optic by its neighbors (Fig. 5.7b).

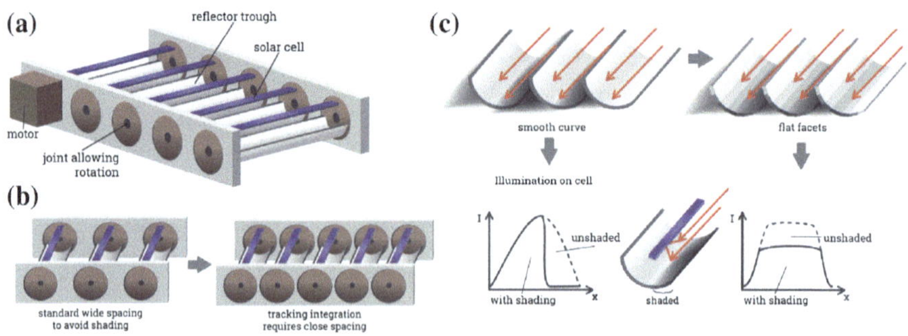

Fig. 5.7 Integrated rotational tracking: **a** the module design uses a line-focus optic (the parabolic trough) and tracks in one dimension by using an integrated motor to rotate the troughs towards the sun. **b** Trackers integrated into a module cannot be spaced to avoid shading. **c** Replacing smooth surfaces with flat-faceted approximations can ensure that cell illumination remains uniform even which the concentrator is partially shaded

In a module that consists of many cells, partial shading of the module is destructive because it limits the current of the whole module to the current of the most shaded cell. If modules are connected to each other in series, the entire string will be limited to the same low current. However, if each shaded element only reflects onto a single cell, there is more flexibility. If the concentrator is a smooth reflective dish, partial shading will cause the light pattern on the cell to be deformed, with one end of the cell being much more brightly illuminated than the other. This can be detrimental to cell performance. In order to rectify this we used a flat-faceted dish that approximates a parabolic curve (Fig. 5.7c). This is a well known solution for improving illumination uniformity on concentrator PV cells [15, 16], and additionally ensures that partial shading of the reflector dish causes only a uniform reduction in the illumination of the cells, rather than deforming the light pattern.

This design could be useful for, perhaps, a 10× concentrator, and has several limitations that will likely prevent it from being useful in higher concentrations. Mechanically, the tracking mechanism is quite straightforward for single-axis tracking but would require more complexity or creativity in order to adapt it into a dual-axis tracker. Second, the trough-based design means that the cell will be more or less thermally isolated, and therefore at anything more than low concentration could become extremely hot, degrading its performance. A final challenge that applies regardless of the concentration factor is the need for flexible internal electrical connection to account for the freely-moving cells.

5.6.1.2 Planar Microtracking

A second concept which has received some attention is to replace the rotational motion of a tracker with a translational motion [17, 18]. In this scenario, sunlight is focused to a spot that moves over the course of the day and year, and the cell is translated to keep it illuminated by the light spot. The light spot can be created by using optical elements that map light from different incidence angles onto different points on a plane. This can be observed most simply in a lens. If collimated light is shined onto a lens, with normal incidence, it will be concentrated to the focal point. If the source is now moved so that the light is incident at a small angle, the light will be directed to a different point, near the focus.

For a single lens, this only works for angles that are close to normal. For larger angles, light will not be focused to the same plane as the focal point. Tracing the image locations for many incidence angles gives a curved surface. This phenomenon is known as Petzval image field curvature. For a single lens this curvature is very strong and prevents any practical microtracking design. In order to flatten this curvature, a number of researchers have used composite optics. The most effective has been to use a refractive optic combined with a reflective optic, illuminating a cell in between the two stages [19] (Fig. 5.8). The module achieves sun tracking by using a linear actuator to shift the position of a transparent plate in the focal plane which holds the solar cells.

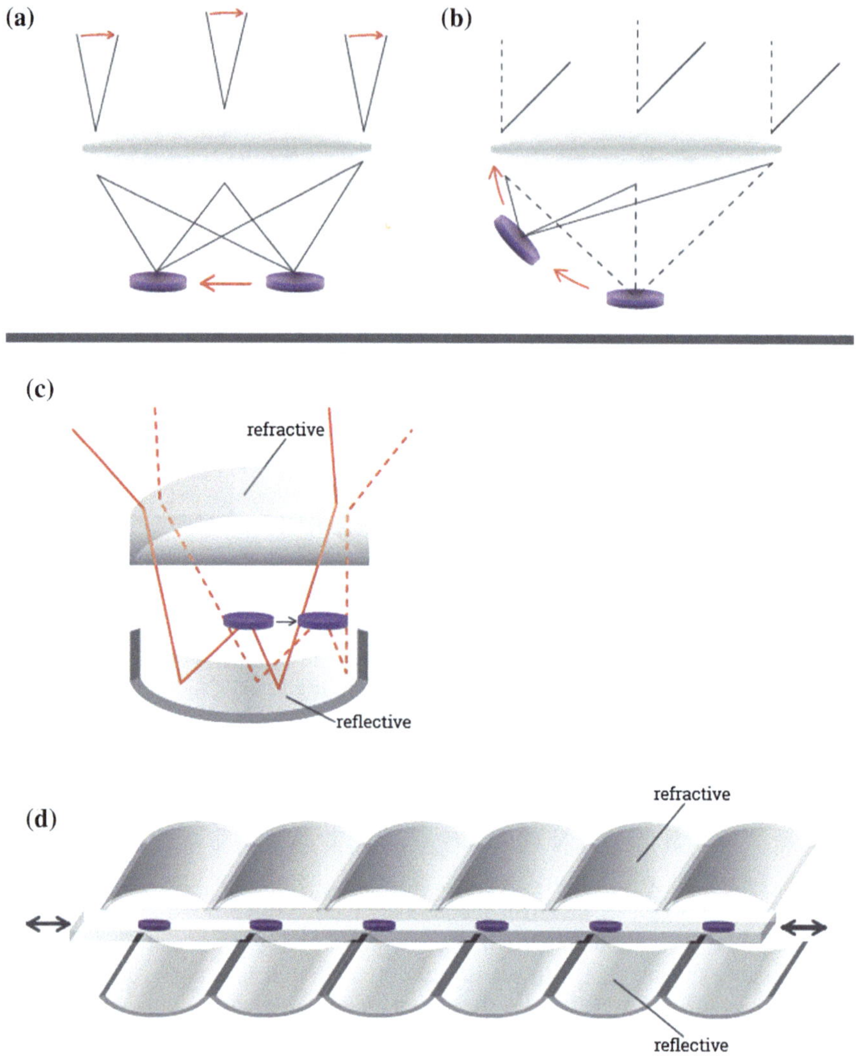

Fig. 5.8 Image field curvature and planar microtracking

There are a few potential benefits to this approach. While the optics are slightly more complex, there is potentially a reduction in mechanical complexity of the tracking system, and possibly as well in energy consumption, from replacing wide-angle rotation of many quasi-independent concentrators with small translations of a single large plate. The tracking can be performed equally well in two axes as in one axis, simply by adding a second actuator. As a result it is straightforward to go to the 100× concentration regime, and possibly higher into the CPV range (500x). In fact it has been shown that planar tracking systems in general can achieve concentrations approaching the etendue limit [18]. The larger number of interfaces could potentially lead to internal optical loss, but this was addressed in tests with index-matching oil that also acts as a lubricant. The thermal isolation of the cells could still pose a challenge although it has been claimed that significant cell heating was not observed in tests.

There are, in each of these proposals, potential complications arising from the mechanical tracking system. These could potentially make tracking integrated solar panels a hard sell, as it has become expected that they are very low-maintenance, although it should be noted that many household appliances get far more wear on their internal machinery than would either of these module concepts. However, whether tracking integration can compete on cost with flat-plate will be the big test.

5.6.1.3 More Advanced Concepts

Other approaches to tracking integration have also been investigated, at the more purely experimental level. The aim is in general to minimize as much as possible the mechanical requirements on the tracker by replacing mechanical tracking altogether with "optical tracking." This uses variations in optical properties of a sun-tracking element, rather than physical movement of any component. Numerous variations have been demonstrated, including electrically-controlled liquid prisms [20], filling and draining an index-matching fluid from a waveguiding optic [21], and fully "reactive" designs that use the change in solar position to directly actuate the tracking behavior [22, 23]. However these remain in the experimental stage.

If concentrator PV has a future, we believe that tracking integration will be central. This comes from the current economic realities and the optical principles discussed here. As economies of scale, and possible material innovations such as those discussed in Chap. 2, continue to drive down the cost of photovoltaic cells, we come closer and closer to a situation where PV cells are extremely cheap, and not significantly higher in cost than commodity materials. This would make it completely impossible to achieve cost savings from CPV, as was once envisioned. The use for concentrators could still be to allow high-cost, high efficiency cells for particular applications, but the value of high efficiency is not always clear, as we have discussed in this chapter. The most obvious advantage, which is the ability to generate more power in a space-constrained setting, is not actually served by today's concentrators, which thanks to the requirements of external mechanical tracking in fact requires a larger total footprint per unit energy output than flat plate PV. As we showed here, only tracking integration can allow full utilization of incident light by a concentrator system. Therefore this may be the single most important element to focus research and development on for those who are serious about rethinking CPV to address real market needs.

But, even using combinations of the many novel technologies that we have described in the last three chapters, the question still remains: does CPV have a future at all? Do these innovations, taken together, indicate a pathway back to relevance for CPV, or are they relics of an earlier era, and a technology whose time has passed?

This question will occupy the next chapter and take us to the conclusion of this book.

References

1. Winston, R. (1974). Principles of solar concentrators of a novel design. *Solar Energy, 16,* 89–95.
2. Winston, R., Miñano, J. C., & Benitez, P. G. (2005). *Non-imaging Optics*. Cambridge: Academic Press.
3. Rabl, A. (1976). Comparison of solar concentrators. *Solar Energy, 18,* 93–111.
4. Shanks, K., Senthilarasu, S., & Mallick, T. K. (2016). Optics for concentrating photovoltaics: Trends, limits and opportunities for materials and design. *Renewable and Sustainable Energy Reviews, 60,* 394–407.
5. Duffie, J. A., & Beckman, W. A. (1980). Solar engineering of thermal processes.
6. Díaz-Dorado, E., Suárez-García, A., Carrillo, C. J., & Cidrás, J. (2011). Optimal distribution for photovoltaic solar trackers to minimize power losses caused by shadows. *Renewable Energy, 36,* 1826–1835.
7. Kerr, A. (2013). *Home Power Magazine*.
8. Philipps, S., Bett, A., Horowitz, K., & Kurtz, S. (2015). *Current status of concentrator photovoltaic (CPV) technology*. CO.: National Renewable Energy Laboratory (NREL), Golden.
9. Bushong, S. (2016). *Solar Power World*.
10. GreentechMedia. (2012). *To Track or Not to Track*. https://www.greentechmedia.com/articles/read/Solar-Balance-of-System-To-Track-or-Not-to-Track-Part-I.
11. Ong, S., Campbell, C., Denholm, P., Margolis, R., & Heath, G. (2013). *Land-use requirements for solar power plants in the United States*. National Renewable Energy Laboratory, Golden, CO 140.
12. Apostoleris, H., Stefancich, M., & Chiesa, M. (2016). Tracking-integrated systems for concentrating photovoltaics. *Nature Energy*, 16018.
13. Apostoleris, H., Leland, J., Chiesa, M., & Stefancich, M. (2016). In *SPIE Optical Engineering + Applications*, (pp. 995507–995506). International Society for Optics and Photonics.
14. Coughenour, B. M., et al. (2014). Dish-based high concentration PV system with Köhler optics. *Optics Express, 22,* A211–A224.
15. Antonini, A., Stefancich, M., Coventry, J., & Parretta, A. (2013). Modelling of compound parabolic concentrators for photovoltaic applications. *International Journal of Optics and Applications, 3,* 40–52.
16. Liu, Z., Lapp, J., & Lipiński, W. (2012). Optical design of a flat-facet solar concentrator. *Solar Energy, 86,* 1962–1966.
17. Hallas, J. M., Karp, J. H., Tremblay, E. J., & Ford, J. E. (2012). In *SPIE Solar Energy + Technology*, (pp. 776904–776907). International Society for Optics and Photonics.
18. Grede, A. J., Price, J. S., & Giebink, N. C. (2016). Fundamental and practical limits of planar tracking solar concentrators. *Optics Express, 24,* A1635–A1646.
19. Price, J. S., Sheng, X., Meulblok, B. M., Rogers, J. A., & Giebink, N. C. (2015). Wide-angle planar microtracking for quasi-static microcell concentrating photovoltaics. *Nature communications, 6.*
20. Cheng, J. (2018). In *APS March Meeting Abstracts*, 1261.

21. DiDomenico, L. D. (2015). Towards doubling solar harvests using wide-angle, broad-band microfluidic beam steering arrays. *Optics Express, 23,* A1398–A1417.
22. Zagolla, V., Tremblay, E., & Moser, C. (2014). Proof of principle demonstration of a self-tracking concentrator. *Optics Express, 22,* A498–A510.
23. Apostoleris, H. N., Chiesa, M., & Stefancich, M. (2016). In *SPIE Optical Engineering + Application*, (pp. 95720A–95726). International Society for Optics and Photonics.

Chapter 6
What Comes Next for CPV?

We have kept returning to the question throughout this book of how we can leverage the developments that we have described in Chaps. 3, 4, 5, to relax the constraints on CPV that were described in Chap. 2, which have prevented it from being competitive with the ultra-cheap flat-plate PV described in Chap. 1. Here we try to provide an answer.

A CPV system is necessarily more complex than a flat-plate module; therefore it will almost unavoidably be more expensive, per unit, to make, as we saw in Chap. 4 . In order to be economically viable, this cost must be compensated by increased power output. However as we showed, this is often not the case for CPV.

High-precision optics with narrow acceptance angles and the need for sun tracking increase costs and decrease output of CPV systems. To address this we have looked at the potential to increase acceptance angles through better optical design, and to rethink tracking, in particular through tracking integration, to make the process less physically demanding and consequently cheaper. However, both approaches have their own limits. At the high concentrations required in today's CPV, the widest possible acceptance angle is only a few degrees, and to our knowledge no device, except for impractically long nonimaging concentrators, have been demonstrated to approach this limit. Meanwhile, practical tracking-integration systems have struggled to reach high concentrations with simultaneously high optical efficiency. Although some promising progress has been made recently it still seems more likely than not that such a system would not produce electricity at a price comparable to flat-plate, and therefore would be limited to niche applications where space constraints dominate.

The only way to relax these constraints further, in a high-efficiency system, is to reduce significantly the concentration factors required. But since these are dictated by the cost of cells, it is necessary first to reduce these costs, probably by a factor of ten or more. This was the point of our first chapter, surveying the attempts to make low-cost multijunction cells. But this push is limited by available materials. To date, as we showed, only the expensive III-Vs have performed suitably in real-life settings as multijunction devices, and, despite much research and hype, a low-cost

H. Apostoleris et al., *Concentrating Photovoltaics (CPV): The Path Ahead*,
Green Energy and Technology, DOI 10.1007/978-3-319-62980-3_6

solution using these materials still seems distant. If achieved, it would be a great breakthrough that would completely overturn many of our conclusions! But such a development is far from certain.

A third approach that has been discussed is that there may, with appropriate design, still be situations where low concentration can be used with high-quality Si to reduce costs. There are some solutions along these lines commercially available. However if our analysis is correct, it seems that these also are bound to be undercut by the ever-dropping cost of flat-plate modules, as the cost of solar cells and the cost of optics are now quite similar!

There is, running through all of this, a more fundamental problem: any new technology—CPV, a new type of cell, etc.—would need to beat the economies of scale of flat-plate PV. It is a question of how far each technology is down its learning curve. Flat plate Si is mature. CPV is less so. Taking this broader view, it is hard to see how a new technology that promises only incremental improvement, and requires huge up-front investment, could catch up with Silicon.

From this same economic perspective, we have considered the possibility that some "shock" could provide an opening for CPV. One possible scenario is the module prices rise sharply in the near future, which would change the economic calculation for prospective installers. This is possible—we see that prices this decade have declined much faster than the learning curve would predict. As a result of this the industry has undergone consolidation as smaller firms have failed. The stage seems set for prices to stabilize as the remaining large players try to recover their losses. But will prices go so high that there would be enough space for entirely new technologies—for that is what would be needed to make CPV widely useful—to break into the mainstream at competitive prices? To us, this seems unlikely.

And so we return to the question, what is left for CPV? In the last three chapters we have gone on a tour through the periphery of CPV—a survey of experimental technologies that never made it to large-scale commercial deployment. All of them have been developed as ways of circumventing the many problems with CPV that have kept it from wide commercial success. But even considering all of these options we are left with the feeling that none of them really is sufficient, at least in the short term, to make it viable. The competition from flat-plate is too strong and the benefits of moving to CPV are not clear enough to justify the substantial effort that would be needed to make the change.

The description that we offered at the beginning of this book holds up: CPV has been beaten at its own game by ultra-cheap silicon PV. If we are serious about furthering the solar revolution, we should not be trying to unseat silicon. We should be trying to complement it. With this guiding principle in mind we propose a paradigm shift.

It is time to break up CPV and reassemble it into something that better suits the needs of the solar age.

6.1 The New CPV

What, we ask ourselves, is the most general purpose of CPV?

It is to use optics to enhance photovoltaic systems.

That this has been narrowed down in recent years to the highly constrained concept of the CPV module is a consequence of the attempt, which we now see as misguided, to create a system analogous to the flat-plate PV module. We have seen how this leads to many complications. But these complications and constraints have led to many innovations that might otherwise not have been made.

In this way our work on CPV that has occurred up to now can be compared to how automobile makers sponsor car racing. It is not the race they care about—it is the technological innovations that are made along the way. The race serves as a "stress test" for developments that may later be broadly applied under everyday conditions.

CPV researchers have been the racecar designers. The technological developments that are made to face the extreme racing conditions can be applied more broadly to make a better consumer car. Now is the time to reap the benefits, and open up the full spectrum of possible applications for our innovations.

In what ways can we reapply our CPV knowledge to make PV, and possibly other systems as well, better? It will be difficult to make it cheaper. We have argued, to a degree, that it is difficult to make it more efficient, but that is not entirely true if we think creatively. We can make it more versatile, with a wider range of functionalties and broader applicability. And we can make it more beautiful—not simply a piece of machinery to be tucked away where most inconspicuous, but a design element, a conversation piece, an artistic statement!

In order to do this we need to fully leverage the capabilities of "unconventional CPV"—the many innovations which we have discussed that were inspired by the heavy constraints of the CPV module paradigm and point the way to new applications.

With this in mind, we suggest simply that the best way forward is to change how we think of the term "CPV" itself. Instead of "concentrator photovoltaics," we will be better served and make better use of the technologies if with think of it as meaning "Concentrators *and* photovoltaics." We call this concept "C+PV," and it refers to any system where optics and photovoltaics are used together to increase functionality or improve utilization of the solar resource. CPV is a special case of this broader class of C+PV technologies. This shift will free from the unnecessary constraints of conventional CPV the many optical innovations that have accompanied its development, and design new systems for this new age that are practical, multifunctional and, if we dare to dream, even beautiful!

C+PV, liberated from the constraints of the CPV module, can open new frontiers in architecture and city planning. Optical design elements integrated into buildings can harvest, produce and redirect light as needed, illuminating dark alleys and corners, shading overheated hallways and streets, produce electricity and heat from conventional solar energy technologies and simultaneously create pleasing visual effects with light-splitting devices designed both for functionality and aesthetics. Homes which reapply CPV elements to enhance natural lighting and provide climate control, in addition to energy production, can be more comfortable and livable, with less energy consumption and environmental impact. And fields unrelated to energy production may find themselves benefited in surprising ways be the repurposing of CPV-derived technology for new uses.

What are some of the things we have learned from CPV that can now be put to work in this new framework?

CPV has "taught" us how to do many things in a cost-effective way that may not have been developed otherwise.

CPV (and CSP) supported the development of ultrahigh-reflectance materials, that have now found additional applications in illumination and consumer electronics.

It has led to innovation in various forms of light splitting that can be applied to make much more efficient use of the solar energy resource, with economical and available technology. Schemes to produce highly compact sun-tracking systems, inspired by CPV, seem likely to find broader use as well.

CPV raised the issue of disposing large thermal loads created by the concentration of sunlight. Some of the solutions to this problem, such as microfluidic cooling [1], later found use in consumer electronics, for example, for thermal management in microprocessors. While these solutions proved too costly for use in commercial systems, the experience of CPV highlighted the critical issue of thermal management in solar cells—another area where light-splitting has found new use.

The entire concept of light-splitting, as we have hinted, is ripe for a new wave of innovation. Spectrum splitting has been heavily investigated for systems of multiple photovoltaic cells, but aside from the III-Vs, there are no cells available to adequately cover the solar spectrum. However the different portions of the spectrum have a range of uses beyond electricity production. For example, long-wave radiation can be used for heating; visible for lighting and agriculture; and the ultraviolet for photocatalysis. In addition, the splitting between direct and diffuse radiation may be valuable, for example, in determining the temperature that can be reached in a thermal system, or in separating "soft" light that is desirable for illumination from "harsh" direct light hat can be directed to other purposes. In Fig. 6.1, we present a schematic of the potential uses for the various parts of the solar spectrum, to guide the reinvention of spectrum splitting to recognize its full potential.

One of the critical insights is that we limit the "PV" portion of the spectrum to the dominant PV technology, Si. There is no point in worrying about other cell technologies that have not reached commercial viability when PV is just one of an array of applications that sunlight can be used for. In its focus on a single use for sunlight—electricity production—CPV treats as "losses" potentially useful features. A broader perspective on spectrum-splitting possibilities avoids such waste, and may ultimately allow the highly efficient utilization of sunlight that has eluded CPV.

This problem of treating useful energy as a loss because it cannot be used by the particular

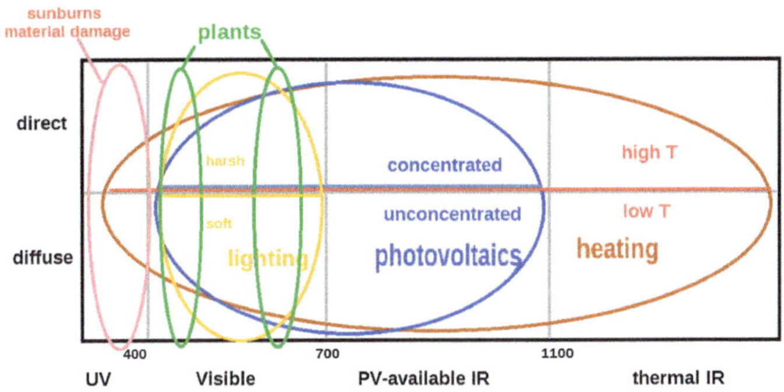

Fig. 6.1 Division of the solar spectrum by potential uses—a guide to reimagining spectrum splitting

technology at hand is a common problem in solar engineering. Consider for example windows in building. Because the purpose of a window is for daylighting, non-visible radiation, or visible radiation in excess of what is pleasant for the human eye, is typically not desired. Partially-reflective window films and smart glasses are designed to reject light that is not used to light the interior. But again, this throws away energy that could be useful. CPV research has presented new possibilities for redirecting this "unwanted" radiation to other uses, for example through wavelength-selective scattering or luminescent waveguides for photovoltaic windows that are

transparent to the visible while directing other wavelengths towards PV for electricity production.

Another window-related application that has been enhanced by cross-talk with CPV is the switchable-transparency material. This has been explored for some time both as a "smart window" that changes its transmissivity depending on conditions, and as an approach to module-integrated sun tracking. One can imagine this also being combined with the previous concept of light splitting in building windows, by using a variable change in scattering characteristics to modify the distribution of light between different applications (for example, between illumination and PV).

Continuing the theme of variable transparency, there is no reason why applications should be limited to the spectral range of sunlight. Reflecting thermal IR radiation (6-10um), for example, is a very effective way of providing thermal insulation. A material that could significantly change transparency in this range could serve as a thin and flexible "smart insulation" that can change depending on the temperature—for example in portable housing or even in clothes.

Funneling sunlight inside of building through the use of roof-mounted light pipes takes advantage both of materials developed for CPV and could be further advanced by the integration of sun tracking and spectral management technologies, also initiated in the context of CPV.

Concepts of CPV can be extended to more specialized applications as well. In some cases this is actually a return to their origins! For example, CPV adopted many optical concepts from astronomy, but recast them in a large scale, low-cost implementation better suited to the economic and physical constraints on PV. The resulting optical devices may provide more practical or economic performance than in their original formulation. For example, if a nonimaging concentrator is returned to its original function of collecting radiation from distant celestial objects, there are a much broader array of devices, which are much more economical and better optimized and understood, than the early CPC that was used for this purpose.

Finally we consider that an energy technology should be measured not only be how much energy it creates, but by how much energy consumption it or its derivatives avoid. In this framework it is especially noteworthy that CPV (and CSP) contributed heavily to artificial illumination. As geometric optics are inherently reversible, CPV research for the ultimate high efficiency concentrator, both benefited from and contributed to the development of high efficiency reflectors geometries and beam shaping methods for LED based illumination systems.

LEDs are, in fact, high divergence and intensity light sources effectively acting as the counterpart of the solar cell in CPV systems when the role of source and target are switched.

This, in turn, contributed to the widespread adoption of LED illumination systems and energy saving dwarfing the direct energy contribution of CPV itself.

Many great developments in science and technology have begun as something completely different than where they ended up, and this seems likely to be the greatest achievement of CPV.

CPV's multiple contributions have, and will continue to trickle down to different applications and fields and, even should CPV fail to become a viable player in the broader PV market, will remain forever as its positive legacy in the future.

Reference

1. Vincenzi, D. *et al.* in *Proceedings of the PV in Europe Conference and Exhibition—From PV technology to Energy Solutions Conference Record.* 7–11.

FSC
www.fsc.org
MIX
Papier | Fördert
gute Waldnutzung
FSC® C083411

Kolibri 360 GmbH
Ferdinand-Jühlke-Straße 7
99095 Erfurt, Deutschland
produktsicherheit@kolibri360.de